2016 SQA Past Papers With Answers

Higher
HUMAN BIOLOGY

2014 Specimen Question Paper, 2015 & 2016 Exams

HODDER GIBSON
AN HACHETTE UK COMPANY

This book contains the official 2014 SQA Specimen Question Paper; 2015 Exam and 2016 Exam for Higher Human Biology, with associated SQA-approved answers modified from the official marking instructions that accompany the paper.

In addition the book contains study skills advice. This advice has been specially commissioned by Hodder Gibson, and has been written by experienced senior teachers and examiners in line with the new Higher for CfE syllabus and assessment outlines. This is not SQA material but has been devised to provide further guidance for Higher examinations.

Hodder Gibson is grateful to the copyright holders, as credited on the final page of the Answer Section, for permission to use their material. Every effort has been made to trace the copyright holders and to obtain their permission for the use of copyright material. Hodder Gibson will be happy to receive information allowing us to rectify any error or omission in future editions.

Hachette UK's policy is to use papers that are natural, renewable and recyclable products and made from wood grown in sustainable forests. The logging and manufacturing processes are expected to conform to the environmental regulations of the country of origin.

Orders: please contact Bookpoint Ltd, 130 Park Drive, Milton Park, Abingdon, Oxon OX14 4SE. Telephone: (44) 01235 827720. Fax: (44) 01235 400454. Lines are open 9.00–5.00, Monday to Saturday, with a 24-hour message answering service. Visit our website at www.hoddereducation.co.uk. Hodder Gibson can be contacted direct on: Tel: 0141 333 4650; Fax: 0141 404 8188; email: hoddergibson@hodder.co.uk

This collection first published in 2016 by
Hodder Gibson, an imprint of Hodder Education,
An Hachette UK Company
211 St Vincent Street
Glasgow G2 5QY

Typeset by Aptara, Inc.

Printed in the UK

A catalogue record for this title is available from the British Library

ISBN: 978-1-4718-9094-9

3 2 1

2017 2016

Introduction

Study Skills – what you need to know to pass exams!

Pause for thought

Many students might skip quickly through a page like this. After all, we all know how to revise. Do you really though?

Think about this:

"IF YOU ALWAYS DO WHAT YOU ALWAYS DO, YOU WILL ALWAYS GET WHAT YOU HAVE ALWAYS GOT."

Do you like the grades you get? Do you want to do better? If you get full marks in your assessment, then that's great! Change nothing! This section is just to help you get that little bit better than you already are.

There are two main parts to the advice on offer here. The first part highlights fairly obvious things but which are also very important. The second part makes suggestions about revision that you might not have thought about but which WILL help you.

Part 1

DOH! It's so obvious but …

Start revising in good time

Don't leave it until the last minute – this will make you panic.

Make a revision timetable that sets out work time AND play time.

Sleep and eat!

Obvious really, and very helpful. Avoid arguments or stressful things too – even games that wind you up. You need to be fit, awake and focused!

Know your place!

Make sure you know exactly **WHEN and WHERE** your exams are.

Know your enemy!

Make sure you know what to expect in the exam.

How is the paper structured?

How much time is there for each question?

What types of question are involved?

Which topics seem to come up time and time again?

Which topics are your strongest and which are your weakest?

Are all topics compulsory or are there choices?

Learn by DOING!

There is no substitute for past papers and practice papers – they are simply essential! Tackling this collection of papers and answers is exactly the right thing to be doing as your exams approach.

Part 2

People learn in different ways. Some like low light, some bright. Some like early morning, some like evening or night. Some prefer warm, some prefer cold. But everyone uses their BRAIN and the brain works when it is active. Passive learning – sitting gazing at notes – is the most INEFFICIENT way to learn anything. Below you will find tips and ideas for making your revision more effective and maybe even more enjoyable. What follows gets your brain active, and active learning works!

Activity 1 – Stop and review

Step 1

When you have done no more than 5 minutes of revision reading STOP!

Step 2

Write a heading in your own words which sums up the topic you have been revising.

Step 3

Write a summary of what you have revised in no more than two sentences. Don't fool yourself by saying, "I know it, but I cannot put it into words". That just means you don't know it well enough. If you cannot write your summary, revise that section again, knowing that you must write a summary at the end of it. Many of you will have notebooks full of blue/black ink writing. Many of the pages will not be especially attractive or memorable so try to liven them up a bit with colour as you are reviewing and rewriting. **This is a great memory aid, and memory is the most important thing.**

Activity 2 – Use technology!

Why should everything be written down? Have you thought about "mental" maps, diagrams, cartoons and colour to help you learn? And rather than write down notes, why not record your revision material?

What about having a text message revision session with friends? Keep in touch with them to find out how and what they are revising and share ideas and questions.

Why not make a video diary where you tell the camera what you are doing, what you think you have learned and what you still have to do? No one has to see or hear it, but the process of having to organise your thoughts in a formal way to explain something is a very important learning practice.

Be sure to make use of electronic files. You could begin to summarise your class notes. Your typing might be slow, but it will get faster and the typed notes will be easier to read than the scribbles in your class notes. Try to add different fonts and colours to make your work stand out. You can easily Google relevant pictures, cartoons and diagrams which you can copy and paste to make your work more attractive and **MEMORABLE**.

Activity 3 – This is it. Do this and you will know lots!

Step 1

In this task you must be very honest with yourself! Find the SQA syllabus for your subject (www.sqa.org.uk). Look at how it is broken down into main topics called MANDATORY knowledge. That means stuff you MUST know.

Step 2

BEFORE you do ANY revision on this topic, write a list of everything that you already know about the subject. It might be quite a long list but you only need to write it once. It shows you all the information that is already in your long-term memory so you know what parts you do not need to revise!

Step 3

Pick a chapter or section from your book or revision notes. Choose a fairly large section or a whole chapter to get the most out of this activity.

With a buddy, use Skype, Facetime, Twitter or any other communication you have, to play the game "If this is the answer, what is the question?". For example, if you are revising Geography and the answer you provide is "meander", your buddy would have to make up a question like "What is the word that describes a feature of a river where it flows slowly and bends often from side to side?".

Make up 10 "answers" based on the content of the chapter or section you are using. Give this to your buddy to solve while you solve theirs.

Step 4

Construct a wordsearch of at least 10 X 10 squares. You can make it as big as you like but keep it realistic. Work together with a group of friends. Many apps allow you to make wordsearch puzzles online. The words and phrases can go in any direction and phrases can be split. Your puzzle must only contain facts linked to the topic you are revising. Your task is to find 10 bits of information to hide in your puzzle, but you must not repeat information that you used in Step 3. DO NOT show where the words are. Fill up empty squares with random letters. Remember to keep a note of where your answers are hidden but do not show your friends. When you have a completed puzzle, exchange it with a friend to solve each other's puzzle.

Step 5

Now make up 10 questions (not "answers" this time) based on the same chapter used in the previous two tasks. Again, you must find NEW information that you have not yet used. Now it's getting hard to find that new information! Again, give your questions to a friend to answer.

Step 6

As you have been doing the puzzles, your brain has been actively searching for new information. Now write a NEW LIST that contains only the new information you have discovered when doing the puzzles. Your new list is the one to look at repeatedly for short bursts over the next few days. Try to remember more and more of it without looking at it. After a few days, you should be able to add words from your second list to your first list as you increase the information in your long-term memory.

FINALLY! Be inspired...

Make a list of different revision ideas and beside each one write **THINGS I HAVE** tried, **THINGS I WILL** try and **THINGS I MIGHT** try. Don't be scared of trying something new.

And remember – "FAIL TO PREPARE AND PREPARE TO FAIL!"

Higher Human Biology

The practice papers in this book give an overall and comprehensive coverage of assessment of **Knowledge** and **Skills of Scientific Inquiry** for the new CfE Higher Human Biology.

We recommend that you refer to Higher Human Biology Course Support Notes pages 9–56 from the SQA website at www.sqa.org.uk. You should note that in your examination only the material included in the Mandatory Course key areas can be examined. Skills of Scientific Enquiry described on pages 59–62 are also examined.

The course

The Higher Human Biology Course consists of two full National Units, which are Human Cells and Physiology, and two half National Units, which are Neurobiology and Health and Communication and Immunology and Public Health. In each of the Units you will be assessed on your ability to demonstrate and apply knowledge of Human Biology and to demonstrate and apply skills of scientific inquiry. Candidates must also complete an Assignment in which they research a topic in biology and write it up as a report. They also take a Course examination.

How the course is graded

To achieve a course award for Higher Human Biology you must pass all four National Unit Assessments which will be assessed by your school or college on a pass or fail basis. The grade you get depends on the following two course assessments, which are set and graded by SQA.

1. An 800–1200 word report based on an Assignment, which is worth 17% of the grade. The Assignment is marked out of 20 marks, with 15 of the marks being for scientific inquiry skills and 5 marks for the application of knowledge.

2. A written course examination is worth the remaining 83% of the grade. The examination is marked out of 100 marks, most of which are for the demonstration and application of knowledge although there are also marks available for skills of scientific inquiry.

This book should help you practise the examination part! To pass Higher Human Biology with a C grade you will need about 50% of the 120 marks available for the Assignment and the Course Examination combined. For a B you will need roughly 60% and, for an A, roughly 70%.

The course examination

The Course Examination is a single question paper in two sections.

- **The first section** is an objective test with 20 multiple choice items for 20 marks.
- **The second section** is a mix of restricted and extended response questions worth between 1 and 9 marks each for a total of 80 marks. The majority of the marks test knowledge with an emphasis on the application of knowledge. The remainder test the application of scientific inquiry, analysis and problem solving skills. There will be a choice offered in the longest questions.

Altogether, there are 100 marks and you will have 2 hours and 30 minutes to complete the paper. The majority of the marks will be straightforward and linked to grade C but some questions are more demanding and are linked to grade A.

General tips and hints

You should download a copy of the Course Assessment Specification (CAS) for Higher Human Biology from the SQA website. This document tells you what can be tested in your examination. It is worth spending some time on this document.

This book contains three practice Higher examination papers. One is the SQA specimen paper and there are two past exam papers. Notice how similar they all are in the way in which they are laid out and the types of question they ask – your own course examination is going to be very similar as well, so the value of the papers is obvious! Each paper can be attempted in its entirety or groups of questions on a particular topic or skill area can be attempted. If you are trying a whole examination paper from this book, give yourself 2 hours and 30 minutes maximum to complete it. The questions in each paper are laid out in Unit order. Make sure that you spend time in using the answer section to mark your own work – it is especially useful if you can get someone to help you with this.

The marking instructions give acceptable answers with alternatives. You could even grade your work on an A–D basis. The following hints and tips are related to examination techniques as well as avoiding common mistakes. Remember that if you hit problems with a question, you should ask your teacher for help.

Section 1

20 multiple-choice items   20 marks

- Answer on the grid.

- Do not spend more than 30 minutes on this section.

- Some individual questions might take longer to answer than others – this is quite normal and make sure you use scrap paper if a calculation or any working is needed.

- Some questions can be answered instantly – again, this is normal.

- Do not leave blanks – complete the grid for each question as you work through.

- Try to answer each question in your head without looking at the options. If your answer is there you are home and dry!

- If you are not certain, choose the answer that seemed most attractive on first reading the answer options.

- If you are guessing, try to eliminate options before making your guess. If you can eliminate three – you are left with the correct answer even if you do not recognise it!

Section 2

Restricted and extended response   80 marks

- Spend about 2 hours on this section.

- Answer on the question paper. Try to write neatly and keep your answers on the support lines if possible – the lines are designed to take the full answer!

- A clue to answer length is the mark allocation – most questions are restricted to 1 mark and the answer can be quite short. If there are 2–4 marks available, your answer will need to be extended and may well have two, three or even four parts.

- The questions are usually laid out in Unit sequence but remember some questions are designed to cover more than one Unit.

- The C-type questions usually start with "State", "Identify", "Give" or "Name" and often need only a word or two in response. They will usually be for one mark each.

- Questions that begin with "Explain" and "Describe" are usually A types and are likely to have more than one part to the full answer. You will usually have to write a sentence or two and there may be two or even three marks available.

- Make sure you read questions over twice before trying to answer – there is often very important information within the question and you are unlikely to be short of time in this examination.

- Using abbreviations like DNA and ATP is fine and the bases of DNA can be given as A, T, G and C. The Higher Human Biology Course Assessment Support Notes will give you the acceptable abbreviations.

- Don't worry that a few questions are in unfamiliar contexts, that's the idea! Just keep calm and read the questions carefully.

- If a question contains a choice, be sure to spend a minute or two making the best choice for you.

- In experimental questions, you must be aware of what variables are, why controls are needed and how reliability and validity might be improved. It is worth spending time on these ideas – they are essential and will come up year after year.

- Some candidates like to use a highlighter pen to help them focus on the essential points of longer questions – this is a great technique.

- Remember that a conclusion can be seen from data, whereas an explanation will usually require you to supply some background knowledge as well.

- Remember to "use values from the graph" when describing graphical information in words if you are asked to do so.

- Plot graphs carefully and join the plot points using a ruler. Include zeros on your scale where appropriate and use the data table headings for the axes labels.

- Look out for graphs with two Y-axes – these need extra special concentration and anyone can make a mistake!

- If there is a space for calculation given – you will very likely need to use it! A calculator is essential.

- The main types of calculation tend to be ratios, averages, percentages and percentage change – make sure you can do these common calculations.

- Answers to calculations will not usually have more than two decimal places.

- Give units in calculation answers if they are not already given in the answer space.

- Do not leave blanks. Always have a go, using the language in the question if you can.

Good luck!

Remember that the rewards for passing Higher Human Biology are well worth it! Your pass will help you get the future you want for yourself. In the exam, be confident in your own ability. If you're not sure how to answer a question, trust your instincts and just give it a go anyway.

Keep calm and don't panic! GOOD LUCK!

National
Qualifications
SPECIMEN ONLY

SQ25/H/02

Human Biology
Section 1 — Questions

Date — Not applicable

Duration — 2 hours and 30 minutes

Instructions for the completion of Section 1 are given on *Page two* of your question and answer booklet SQ25/H/02.

Record your answers on the answer grid on *Page three* of your question and answer booklet.

Before leaving the examination room you must give your question and answer booklet to the Invigilator; if you do not you may lose all the marks for this paper.

SECTION 1 — 20 marks

Attempt ALL questions

1. The diagram below shows two chromosomes, M and N, before and after a chromosomal mutation.

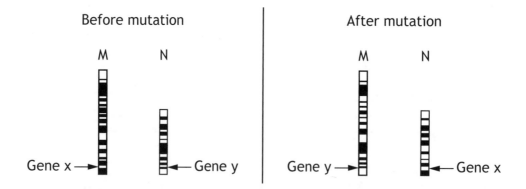

The form of mutation that has taken place is a

A   translocation

B   duplication

C   insertion

D   deletion.

2. Amplification of DNA by PCR commences with 1000 DNA molecules in the reaction tube. How many DNA molecules would be present after four cycles of PCR?

A   4000

B   8000

C   16000

D   32000

3. Which of the following statements about slow twitch muscle fibres is correct?

A   They cannot sustain contractions for as long as fast twitch muscle fibres.

B   They have many more mitochondria than fast twitch muscle fibres.

C   They are better for activities like weightlifting and sprinting than fast twitch muscle fibres.

D   They store fuel mainly as glycogen while fast twitch muscle fibres store fuel as fat.

4. The table below contains information about four semen samples.

	Semen sample			
	A	B	C	D
Number of sperm in sample (millions/cm^3)	40	30	20	60
Active sperm (%)	50	60	75	40
Abnormal sperm (%)	30	65	10	70

Which semen sample has the highest number of active sperm?

5. In which of the following situations might a fetus be at risk from Rhesus antibodies produced by the mother?

	Father	Mother
A	Rhesus positive	Rhesus negative
B	Rhesus positive	Rhesus positive
C	Rhesus negative	Rhesus negative
D	Rhesus negative	Rhesus positive

6. The family tree below shows the pattern of inheritance of a genetic condition.

Unaffected female   x   Unaffected male

Affected female

The allele responsible for this condition is both

A   sex-linked and recessive

B   sex-linked and dominant

C   autosomal and recessive

D   autosomal and dominant.

7. The graph below shows the growth, in length, of a human fetus.

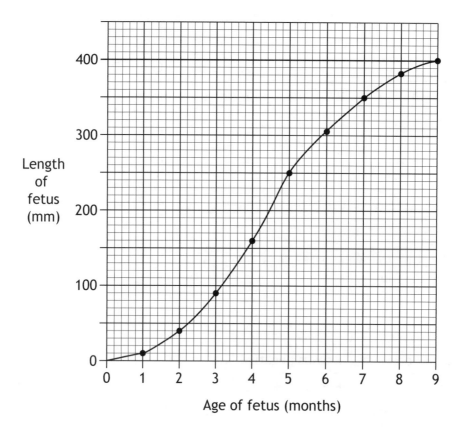

Age of fetus (months)

What is the percentage increase in length of the fetus during the final four months of pregnancy?

A 33·3%

B 60·0%

C 62·5%

D 150·0%

8. Cystic fibrosis is a genetic condition caused by an allele that is not sex-linked.

A child is born with cystic fibrosis despite neither parent having the condition.

The parents are going to have a second child.

What is the percentage chance this child will have cystic fibrosis?

A 75%

B 67%

C 50%

D 25%

9. The duration of the stages in an individual's cardiac cycle are shown in the table below.

Stage	Duration (s)
Diastole	0·4
Atrial systole	0·1
Ventricular systole	0·3

What is the heart rate of this individual?

A   48 beats per minute

B   75 beats per minute

C   80 beats per minute

D   150 beats per minute

10. The diagram below shows a cross-section of the heart.

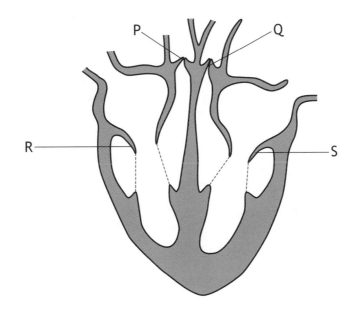

Which of the following statements describes the movement of the valves during ventricular systole?

A   Valves P and Q open and valves R and S close

B   Valves P and R open and valves Q and S close

C   Valves P and Q close and valves R and S open

D   Valves P and R close and valves Q and S open

11. Which of the following statements about lipoprotein is correct?

   A   LDL transports cholesterol from body cells to the heart

   B   LDL transports cholesterol from body cells to the liver

   C   HDL transports cholesterol from body cells to the heart

   D   HDL transports cholesterol from body cells to the liver

12. The graphs below contain information about the population of Britain.

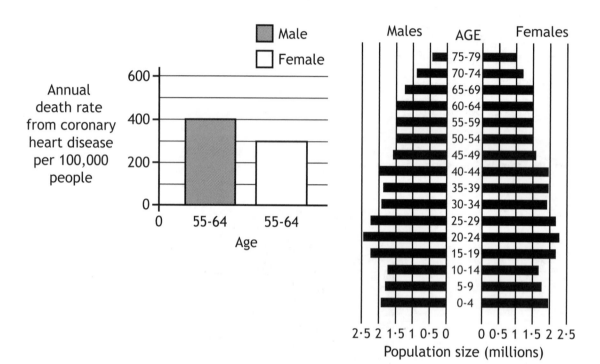

   The number of British women between 55 and 64 years of age who die from coronary heart disease annually is

   A   300

   B   4500

   C   9000

   D   21000.

13. The transformation of information into a form that memory can accept is called

   A   shaping

   B   retrieval

   C   encoding

   D   storage.

14. The diagram below shows a test on a man who had a damaged corpus callosum.

This meant that he could no longer transfer information between his right and left cerebral hemispheres.

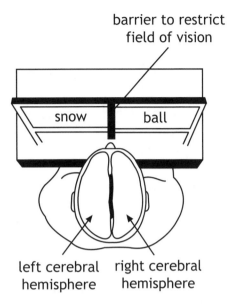

Some of the functions of each hemisphere are described in the table below.

Left cerebral hemisphere	Right cerebral hemisphere
processes information from right eye	processes information from left eye
controls language production	controls spatial task co-ordination

The man was asked to look straight ahead and then the words "snow" and "ball" were flashed briefly on the screen as shown.

What would the man say that he had just seen?

A   Snow

B   Ball

C   Snowball

D   Nothing

15. Which of the following statements about the action of recreational drugs on brain neurochemistry is correct?

A   Desensitisation results from an increase in the number of neurotransmitter receptors due to the use of drugs that are agonists

B   Desensitisation results from an increase in the number of neurotransmitter receptors due to the use of drugs that are antagonists

C   Sensitisation results from an increase in the number of neurotransmitter receptors due to the use of drugs that are agonists

D   Sensitisation results from an increase in the number of neurotransmitter receptors due to the use of drugs that are antagonists

16. An investigation was carried out to determine how long it takes students to learn to run a finger maze.

A blindfolded student was allowed to run the maze on ten occasions.

The results are given in the table below.

Trial	Time (s)
1	23
2	20
3	26
4	12
5	18
6	10
7	6
8	7
9	6
10	6

Which of the following changes to the investigation would make the results more reliable?

A   Allowing other students to try to run the maze ten times.

B   Allowing the same student some additional trials on the same maze.

C   Changing the shape of the maze and allowing the same student to repeat ten trials.

D   Recording the times to one decimal place.

17. Which of the following is not part of the inflammatory response?

    A   Vasodilation

    B   Release of histamine

    C   Production of antibodies

    D   Increased capillary permeability

18. The diagram below represents clonal selection in lymphocytes.

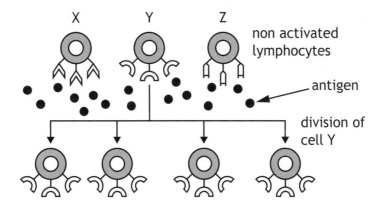

    What stimulates the division of cell Y?

    A   The presence of lymphocytes X and Z

    B   The presence of an antigen in the blood

    C   The binding of antibodies to receptors on the cell membrane

    D   The binding of antigens to receptors on the cell membrane

19. Two groups of subjects were used when carrying out clinical trials of a vaccine.
    One group was given the vaccine while the other group was given a placebo.
    The purpose of the placebo was to

    A   reduce experimental error

    B   ensure a valid comparison can be made

    C   allow a statistical analysis of the results to be made

    D   ensure that researchers are unaware who has been vaccinated.

20. The table below contains data about a worldwide infection in 2009.

	Number of adults	Number of children
Had this infection at the start of 2009	$30·8 \times 10^6$	$2·5 \times 10^6$
Contracted this infection during 2009	$2·2 \times 10^6$	$0·4 \times 10^6$
Died from this infection during 2009	$1·6 \times 10^6$	$0·2 \times 10^6$

How many people in the world had this infection at the start of 2010?

A $35·9 \times 10^6$

B $34·1 \times 10^6$

C $33·3 \times 10^6$

D $31·5 \times 10^6$

[END OF SECTION 1. NOW ATTEMPT THE QUESTIONS IN SECTION 2 OF YOUR QUESTION AND ANSWER BOOKLET.]

Mark

SQ25/H/01

Human Biology
Section 1 — Answer Grid
and Section 2

National
Qualifications
SPECIMEN ONLY

Date — Not applicable

Duration — 2 hours and 30 minutes

Fill in these boxes and read what is printed below.

Full name of centre

Town

Forename(s)

Surname

Number of seat

Date of birth

Day	Month	Year
D D	M M	Y Y

Scottish candidate number

Total marks — 100

SECTION 1 — 20 marks

Attempt ALL questions.

Instructions for completion of Section 1 are given on *Page two*.

SECTION 2 — 80 marks

Attempt ALL questions.

Write your answers in the spaces provided. Additional space for answers and rough work is provided at the end of this booklet. If you use this space, write clearly the number of the question you are attempting. Any rough work must be written in this booklet. You should score through your rough work when you have written your fair copy.

Use **blue** or **black** ink.

Before leaving the examination room you must give this booklet to the Invigilator; if you do not, you may lose all the marks for this paper.

SECTION 1— 20 marks

The questions for Section 1 are contained in the question paper SQ25/H/02.
Read these and record your answers on the answer grid on *Page three* opposite.
Do NOT use gel pens.

1. The answer to each question is **either** A, B, C or D. Decide what your answer is, then fill in the appropriate bubble (see sample question below).

2. There is **only one correct** answer to each question.

3. Any rough working should be done on the additional space for answers and rough work at the end of this booklet.

Sample Question

The digestive enzyme pepsin is most active in the

    A   mouth

    B   stomach

    C   duodenum

    D   pancreas.

The correct answer is **B**—stomach. The answer **B** bubble has been clearly filled in (see below).

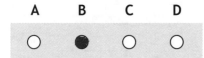

Changing an answer

If you decide to change your answer, cancel your first answer by putting a cross through it (see below) and fill in the answer you want. The answer below has been changed to **D**.

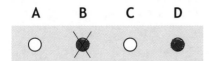

If you then decide to change back to an answer you have already scored out, put a tick (✓) to the **right** of the answer you want, as shown below:

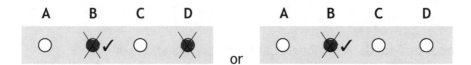

    or

SECTION 1 — Answer Grid

	A	B	C	D
1	○	○	○	○
2	○	○	○	○
3	○	○	○	○
4	○	○	○	○
5	○	○	○	○
6	○	○	○	○
7	○	○	○	○
8	○	○	○	○
9	○	○	○	○
10	○	○	○	○
11	○	○	○	○
12	○	○	○	○
13	○	○	○	○
14	○	○	○	○
15	○	○	○	○
16	○	○	○	○
17	○	○	○	○
18	○	○	○	○
19	○	○	○	○
20	○	○	○	○

MARKS | DO NOT WRITE IN THIS MARGIN

SECTION 2 — 80 marks

Attempt ALL questions

Note that question 14 contains a choice.

1. The human body contains many specialised cells, all of which have developed from stem cells in the early embryo.

Nerve cells

Liver cells

Cardiac muscle cells

(a) Name the process by which a stem cell develops into a specialised body cell and explain how this process occurs.   **2**

Process _____

Explanation _____

_____

(b) Both germline and somatic cells retain the ability to divide.

    (i) State the type of cell division that only occurs in germline cells.   **1**

    _____

    (ii) Explain why mutations in germline cells are potentially more serious than mutations in somatic cells.   **1**

    _____

    _____

(c) A company has developed a drug that could be used to treat the symptoms of an inherited disease. Before proceeding to clinical trials using volunteers, the company decides to carry out additional tests in the laboratory using stem cells.

    Describe one ethical consideration that might have influenced this decision to use stem cells.   **1**

    _____

    _____

MARKS | DO NOT WRITE IN THIS MARGIN

2. The diagram below shows stages in the synthesis of a polypeptide.

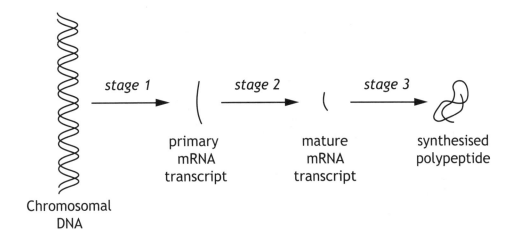

(a) Name the enzyme that catalyses stage 1 of this process.

1

_____

(b) Name stage 3 and state the exact location where it occurs within a cell.

1

Name _____

Location _____

(c) (i) Explain why the primary mRNA transcript is so much shorter than chromosomal DNA.

1

_____

_____

(ii) Explain why the mature mRNA transcript is shorter than the primary mRNA transcript.

1

_____

_____

MARKS | DO NOT WRITE IN THIS MARGIN

3. An experiment was carried out to investigate the effect of substrate concentration on the production of an end-product in an enzyme controlled reaction.

The enzyme urease was used which breaks down urea into ammonia.

$$\text{urea} \xrightarrow{\text{urease}} \text{ammonia}$$

Urease and urea solutions were mixed together and added to test tubes containing agar jelly as shown in the diagram below.

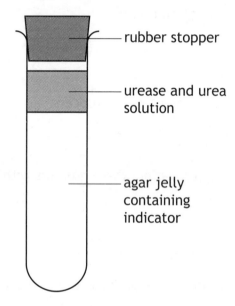

— rubber stopper

— urease and urea solution

— agar jelly containing indicator

Five different concentrations of urea solution were added.

During the reaction the ammonia produced diffused through the agar jelly changing the indicator from yellow to blue.

The length of the agar jelly stained blue was measured after the experiment had been allowed to run for 48 hours.

The results of the experiment are shown in the table below.

Urea concentration added (molar)	Average length of agar jelly stained blue (mm)
0·03	2
0·06	4
0·13	8
0·25	16
0·50	32

MARKS | DO NOT WRITE IN THIS MARGIN

3. **(continued)**

(a) Plot a line graph to illustrate the results of the experiment.

(Additional graph paper, if required, can be found on *Page twenty-six*)   2

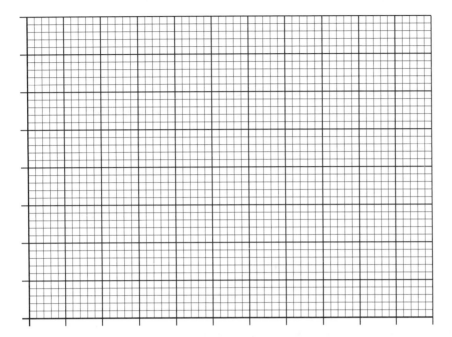

(b)   (i) Name **one** variable that should be controlled when setting up this experiment.   1

_____

(ii) Name **one** variable that should be kept constant during the 48 hours of this experiment.   1

_____

(c) Give the feature of this experiment that makes the results reliable.   1

_____

_____

(d) Explain why the test tubes were left for 48 hours before the results were obtained.   1

_____

_____

MARKS | DO NOT WRITE IN THIS MARGIN

3. **(continued)**

(e) State **one** conclusion that can be drawn from the results of this experiment.

_____

_____

1

(f) Using the **information in the table**, predict the length of agar jelly that would have been stained blue if a 0·75 molar urea solution had been used in the experiment.

Space for calculation

_____ mm

1

(g) Thiourea is a competitive inhibitor of urease.

In another experiment, a test tube of agar jelly was set up containing the urease solution, 0·5 molar urea solution and thiourea.

After 48 hours only 7mm of agar jelly had turned blue.

(i) Explain why less agar jelly turned blue in this experiment than in the first experiment, which also used a 0·5 molar urea solution.

_____

_____

1

(ii) Suggest why 7mm of agar jelly turned blue in this experiment.

_____

_____

1

MARKS | DO NOT WRITE IN THIS MARGIN

4. The diagram below represents the glycolysis stage of respiration in a muscle cell.

Phase 1   *Phase 2*

glucose ⟶ intermediate compounds ⟶ pyruvate

(a) Phase 1 is the energy investment stage of glycolysis while phase 2 is the energy pay-off stage of glycolysis.

Describe what happens during the energy investment and energy pay-off phases of glycolysis.

2

Energy investment phase _____

_____

Energy pay-off phase _____

_____

(b) Once pyruvate has been formed it can be converted into two different compounds, depending on the conditions.

Name one of these compounds and state under what conditions it would be produced.

2

_____

_____

(c) Many athletes take creatine supplements to improve their sporting performance.

State whether sprinters or marathon runners would gain the greatest benefit from taking creatine and give a reason for your choice.

1

Athlete _____

Reason _____

_____

MARKS | DO NOT WRITE IN THIS MARGIN

5. Sickle cell disease is an autosomal blood disorder in which a faulty form of haemoglobin, called haemoglobin S, is produced. This protein is an inefficient carrier of oxygen.

The allele for normal haemoglobin (H) is incompletely dominant to the allele for haemoglobin S (S).

Heterozygous individuals (HS) suffer from a milder condition called sickle cell trait.

The pedigree chart below shows the incidence of these conditions in three generations of a family.

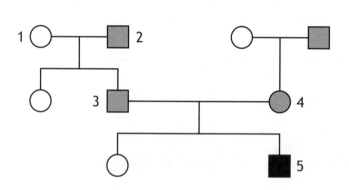

- ▨ male with sickle cell trait
- ■ male with sickle cell disease
- ○ unaffected female
- ⬤ female with sickle cell trait

(a) State the genotype of individual 5.   1

_____

(b) Individuals 3 and 4 go on to have a 3rd child.

State the percentage chance that this child will have the same genotype as the parents.   1

Space for calculation

_____ %

(c) Sickle cell disease is caused by a substitution mutation in the gene that codes for haemoglobin.

(i) Describe how this form of mutation affects the structure of the gene.   1

_____

_____

(ii) Explain how this might change the structure of a protein such as haemoglobin.   1

_____

_____

MARKS | DO NOT WRITE IN THIS MARGIN

5. **(continued)**

(d) During IVF treatment, it is possible to detect single gene disorders in fertilised eggs before they are implanted into the mother.

Give the term that describes this procedure.   **1**

_____

(e) It has been discovered that the gene that codes for fetal haemoglobin is unaffected by the substitution mutation that causes sickle cell disease.

This gene is "switched off" at birth.

Use this information to suggest how a drug designed to treat sickle cell disease in young children could function.   **1**

_____

_____

_____

MARKS | DO NOT WRITE IN THIS MARGIN

6. The diagram below represents a section through an artery.

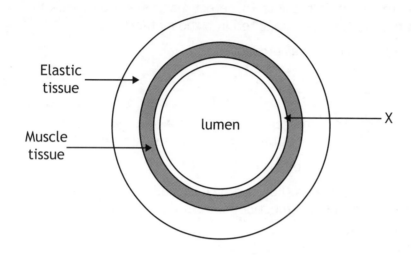

(a) Describe how the presence of muscle tissue in the artery wall helps to control the flow of blood around the body.

1

_____

_____

(b) Describe how an atheroma forming under layer X may lead to the formation of a blood clot and state the possible effects of this.

5

Space for answer

MARKS | DO NOT WRITE IN THIS MARGIN

7. The graph below shows how an individual's heart rate and stroke volume changed as their oxygen uptake increased during exercise.

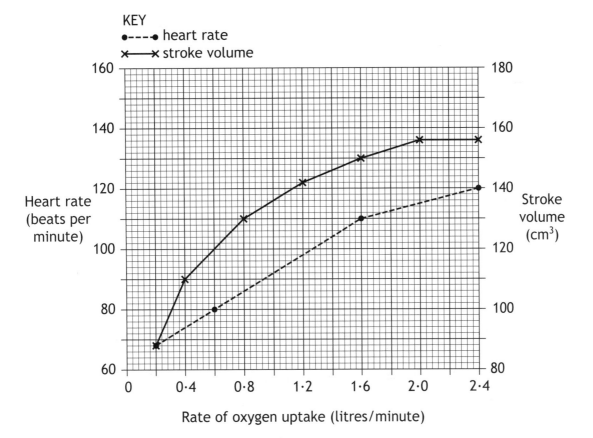

KEY
•----• heart rate
✗——✗ stroke volume

(a) (i) State the individual's heart rate when the rate of oxygen uptake was 1·2 litres/minute.   1

_____

(ii) Using data from the graph, describe how the stroke volume changed as oxygen uptake increased.   1

_____

_____

(iii) State the stroke volume when the heart rate was 110 beats per minute.   1

_____ cm³

MARKS | DO NOT WRITE IN THIS MARGIN

7. (continued)

(b) Calculate the cardiac output when the rate of oxygen uptake was 2·4 litres per minute.

Space for calculation

_____ litres/min

1

(c) (i) When the individual's blood pressure was measured an hour after exercise, a reading of 140/90 mm/Hg was recorded.

Explain why two figures are given for a blood pressure reading.

_____

_____

1

(ii) The individual was diagnosed as having high blood pressure.

One of the effects of this was that their ankles regularly swelled up due to a build-up of tissue fluid.

Explain the link between high blood pressure and the build-up of tissue fluid.

_____

_____

_____

_____

2

MARKS | DO NOT WRITE IN THIS MARGIN

8. The graph below shows changes in blood glucose concentration in a diabetic and a non-diabetic individual after each had consumed a glucose drink.

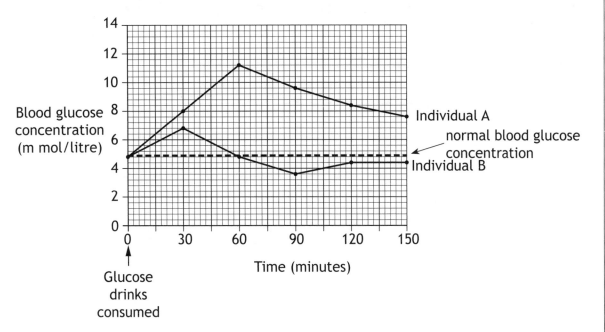

(a) (i) Choose **one** individual, A or B and indicate whether the individual is diabetic or non-diabetic.

Individual _____

Diabetic ☐   Non-diabetic ☐

Using evidence from the graph, justify your choice.

1

_____

_____

(ii) Using data from the graph, describe the changes that occurred in the blood glucose concentration of individual A after consuming the glucose drink.

2

_____

_____

_____

_____

MARKS | DO NOT WRITE IN THIS MARGIN

8.   (continued)

(b)   Describe the role of insulin in the development of type 1 and type 2 diabetes.

2

Type 1 _____

_____

Type 2 _____

_____

MARKS | DO NOT WRITE IN THIS MARGIN

9. The graph below shows obesity data for a European country in 2003 and 2012.

Individuals are described as obese if they have a body mass index (BMI) of 30 or greater.

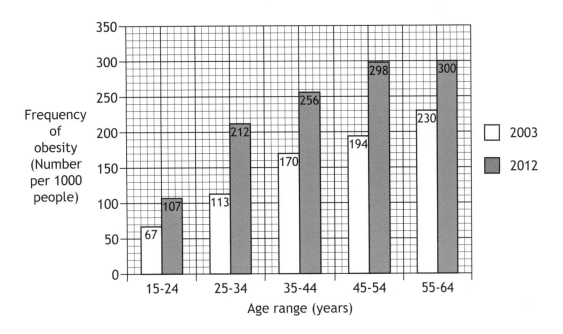

(a) (i) Describe **two** general trends shown in the graph.   2

1 _____

_____

2 _____

_____

(ii) In 2012 the number of people in this country aged 35 to 44 was 6 million.

Calculate how many people aged 35 to 44 were obese.   1

Space for calculation

Number of people _____

(b) State one piece of advice that an obese individual would be given to adapt their diet or lifestyle in order to avoid long-term health problems.   1

_____

_____

MARKS | DO NOT WRITE IN THIS MARGIN

10. A student carried out an investigation into the effect of age on learning ability.

Eight children from three different age groups were each given five attempts to complete a twenty-piece jigsaw puzzle.

The fastest times that they achieved are shown in the table below.

	Fastest time achieved (seconds)		
	8-year-olds	12-year-olds	16-year-olds
	123	97	99
	98	68	74
	111	75	62
	138	112	67
	87	93	84
	136	83	101
	79	75	58
	120	81	55
average	111·5		75·0

(a) Calculate the average fastest time achieved by the 12 year-old children and write your answer in the table above.

Space for calculation

1

(b) Describe **two** additional variables that would have to be kept constant to ensure a valid comparison could be made between the three groups of children.

2

Variable 1 _____

_____

Variable 2 _____

_____

MARKS

DO NOT WRITE IN THIS MARGIN

10. **(continued)**

(c) State a conclusion that can be drawn from the results of this investigation.

1

_____

_____

(d) (i) Explain why the first attempt to complete the puzzle was always slower than the fifth attempt, no matter the age of the child.

1

_____

_____

(ii) Suggest why some children did not produce their fastest time on their fifth attempt.

1

_____

_____

(e) Suggest how the student could adapt the investigation to demonstrate social facilitation.

1

_____

_____

MARKS | DO NOT WRITE IN THIS MARGIN

11. The graph below shows the number of whooping cough cases over a 65 year period in a country.

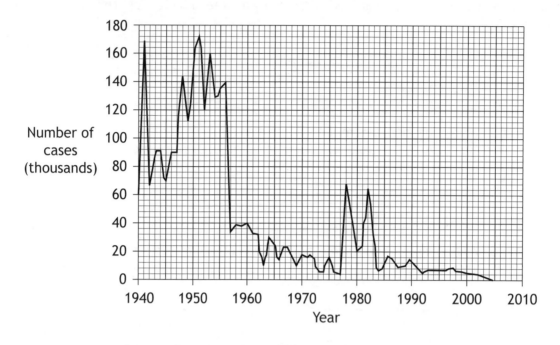

(a) (i) Using information from the graph, state the year in which a vaccine for whooping cough was introduced.　　1

Year _____

(ii) Suggest a reason for the unexpected increase in the number of cases of whooping cough in 1977.　　1

_____

_____

(b) The number of cases of whooping cough decreases to a very low level after 2000 because of herd immunity.

Explain what is meant by the term "herd immunity".　　2

_____

_____

_____

_____

12. The diagrams below contain information about the causes of death and survival rates in two countries in 2010.

Figure 1 - Causes of death in countries A and B during 2010

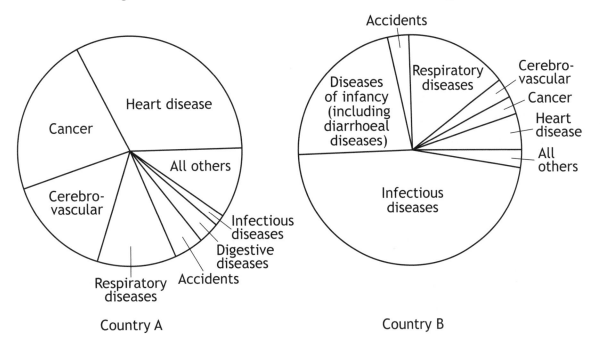

Country A   Country B

Figure 2 - Percentage survival rates in countries A and B in 2010

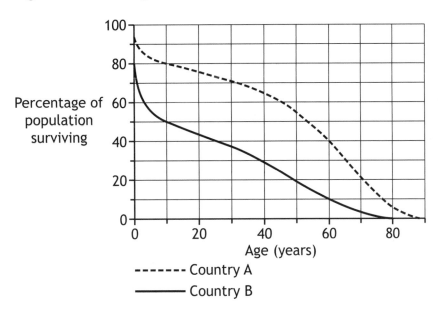

(a) (i) Use information from **Figure 2** to explain the lower incidence of heart disease in Country B.   1

_____

_____

MARKS | DO NOT WRITE IN THIS MARGIN

12.   (a)   (continued)

(ii)   Give an example of how diseases of infancy can be reduced in Country B through community responsibility, other than by vaccination programmes.

1

_____

_____

(b)   (i)   Calculate the percentage of the population of Country A that die before the age of 10.

1

Space for calculation

_____ %

(ii)   In 1950 three million babies were born in Country B.

Calculate how many of these individuals were still alive in 2010, assuming no migration occurred.

1

Space for calculation

_____

MARKS | DO NOT WRITE IN THIS MARGIN

13. Pulmonary tuberculosis (TB) is an infectious disease of the lungs caused by a bacterium.

This bacterium can also damage other organs in the body. When this happens it is called non-pulmonary TB.

The table below shows the number of reported cases of pulmonary and non-pulmonary TB in Scotland between 1981 and 2006.

Year	Number of cases of pulmonary TB	Number of cases of non-pulmonary TB
1981	659	140
1986	500	178
1991	452	97
1996	408	102
2001	275	125
2006	255	153

(a) Suggest how pulmonary TB is transmitted between individuals.   1

(b) (i) In which 5 year period was the greatest decrease in the total number of cases of TB?   1

Space for calculation

(ii) Suggest a reason for this decrease.   1

(iii) Compare the trend in the number of cases of pulmonary TB with that of non-pulmonary TB between 1991 and 2006.   1

MARKS | DO NOT WRITE IN THIS MARGIN

13. (b) (continued)

(iv) Calculate, as a simple whole number ratio, the number of cases of pulmonary TB compared to non-pulmonary TB in 2001.

Space for calculation

1

_____ : _____
pulmonary TB   non-pulmonary TB

(c) Non-pulmonary TB is often associated with HIV infection.

Suggest a reason for this association.

1

_____

_____

MARKS | DO NOT WRITE IN THIS MARGIN

14. Answer **either** A **or** B in the space below.

A   Describe the structure and function of the autonomic nervous system.   **7**

OR

B   Describe the function and mechanism of neurotransmitter action at the synapse.   **7**

[END OF SPECIMEN QUESTION PAPER]

ADDITIONAL SPACE FOR ANSWERS AND ROUGH WORK

Additional Graph for Question 3 (a)

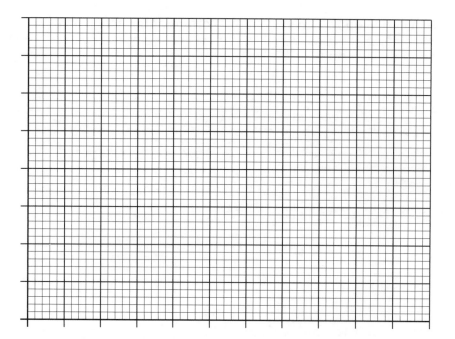

ADDITIONAL SPACE FOR ANSWERS AND ROUGH WORK

MARKS | DO NOT WRITE IN THIS MARGIN

ADDITIONAL SPACE FOR ANSWERS AND ROUGH WORK

ADDITIONAL SPACE FOR ANSWERS AND ROUGH WORK

Page twenty-eight

HIGHER

2015

National Qualifications 2015

X740/76/02

Human Biology
Section 1 — Questions

WEDNESDAY, 13 MAY

1:00 PM – 3:30 PM

Instructions for the completion of Section 1 are given on *Page two* of your question and answer booklet.

Record your answers on the answer grid on *Page three* of your question and answer booklet.

Before leaving the examination room you must give your question and answer booklet to the Invigilator; if you do not you may lose all the marks for this paper.

SECTION 1 — 20 marks

Attempt ALL questions

1. The diagram below shows an enzyme-catalysed reaction taking place in the presence of an inhibitor.

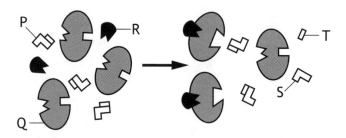

Which line in the table below identifies the molecules in the reaction?

	Inhibitor	Substrate	Product
A	P	R	S
B	Q	P	S
C	R	P	T
D	R	Q	T

2. A primary transcript is a strand of

A   RNA comprising just exons

B   DNA comprising just exons

C   RNA comprising introns and exons

D   DNA comprising introns and exons.

3. The diagram below can be used to identify amino acids coded for by mRNA codons.

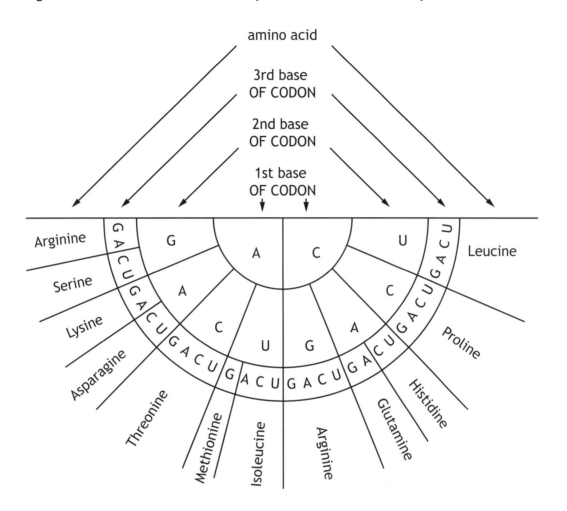

How many different amino acids are coded for by the following mRNA strand?

A U G C C A A C U C C U A G A C G A A U A

A   4

B   5

C   6

D   7

[Turn over

4. The following are descriptions of three single gene mutations.

    Description 1: exon-intron codons are created or destroyed

    Description 2: one amino acid codon is replaced with another

    Description 3: one amino acid codon is replaced with a stop codon

    Which line in the table below matches the descriptions with the correct gene mutation?

	Gene mutation		
	Missense	Nonsense	Splice site
A	1	2	3
B	1	3	2
C	2	1	3
D	2	3	1

5. DNA profiling may be used in criminal investigations.

    During this procedure DNA is cut into fragments by two different enzymes. Each enzyme cuts DNA at a specific point.

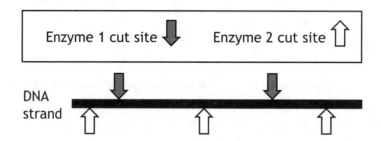

    Which line in the table below gives the correct number of DNA fragments produced from this DNA strand?

	Number of fragments produced using		
	enzyme 1 only	enzyme 2 only	enzymes 1 and 2
A	2	3	5
B	2	3	6
C	3	4	7
D	3	4	6

6. The graph below shows the changes to the concentrations of substrate and product during an enzyme-controlled reaction.

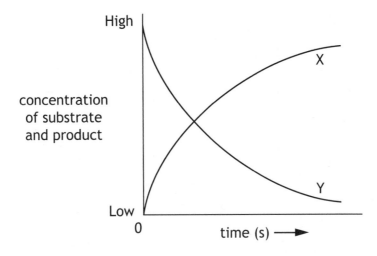

Which line in the table below identifies the substrate, product and the change in the rate of the reaction during the process?

	Substrate	Product	Rate of reaction
A	X	Y	increasing
B	X	Y	decreasing
C	Y	X	increasing
D	Y	X	decreasing

[Turn over

7. The graph below shows the rate of potassium uptake and glucose breakdown by muscle tissue in solutions of different oxygen concentrations.

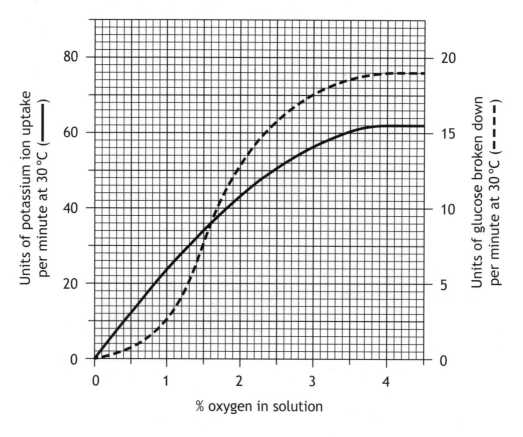

% oxygen in solution

How much glucose is broken down per minute when the oxygen concentration is 1%?

A   2·5 units

B   6 units

C   10 units

D   24 units

8. A 40 g serving of a breakfast cereal contains 2 mg of iron. Only 25% of this iron is absorbed into the bloodstream.

If a pregnant woman requires a daily uptake of 6 mg of iron, how much cereal would she have to eat each day to meet this requirement?

A   60 g

B   120 g

C   240 g

D   480 g

9. The diagram below shows a section through part of the testes.

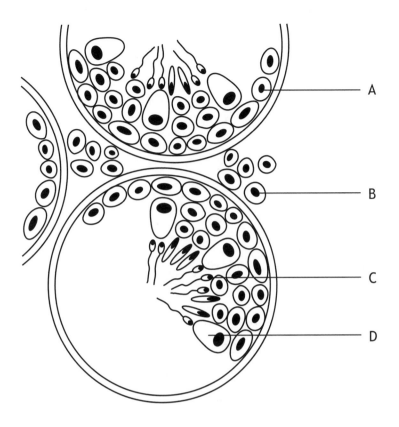

Which cells produce testosterone?

10. The table below shows some genotypes and phenotypes associated with forms of sickle-cell anaemia.

Genotype	Phenotype
AA	unaffected
AS	sickle-cell trait
SS	acute sickle-cell anaemia

A woman with sickle-cell trait and an unaffected man have a child together.

What are the chances that their child will have acute sickle-cell anaemia?

A   None

B   1 in 1

C   1 in 2

D   1 in 4

[Turn over

11. The events leading to formation of a blood clot are listed below.

1. Clotting factors are released.
2. An insoluble meshwork forms.
3. Fibrinogen is converted to fibrin.
4. Prothrombin is converted to thrombin.

The correct sequence of these events is

A   4, 2, 3, 1

B   1, 4, 3, 2

C   1, 3, 4, 2

D   4, 3, 1, 2

12. Which of the following statements describes the role of lipoprotein in the transport and elimination of excess cholesterol?

A   Low density lipoprotein transports excess cholesterol from the liver to the body cells.

B   Low density lipoprotein transports excess cholesterol from the body cells to the liver.

C   High density lipoprotein transports excess cholesterol from the liver to the body cells.

D   High density lipoprotein transports excess cholesterol from the body cells to the liver.

13. Which of the following describes typical features of Type 1 diabetes?

	Feature of Type 1 diabetes	
A	occurs in childhood	cells unable to produce insulin
B	develops later in life	cells unable to produce insulin
C	occurs in childhood	cells less sensitive to insulin
D	develops later in life	cells less sensitive to insulin

14. The following are types of neural pathways.

    1. Diverging
    2. Converging
    3. Reverberating

    Which of these pathways involve nerve impulses being sent back through a circuit of neurons?

    A   3 only

    B   1 and 2 only

    C   1 and 3 only

    D   1, 2 and 3

15. After drinking, alcohol is removed from the blood at a constant rate.

    The table below shows the average time it takes to remove different alcohol concentrations from the blood.

Blood alcohol concentration (mg/100 cm^3)	Removal time (hours)
16	1·0
50	3·125
80	5·0
100	6·25
160	10·0
200	12·5

    The legal maximum blood alcohol concentration for driving in some regions of the UK is 80 mg/100 cm^3.

    Predict how long it would take before a person with a blood alcohol concentration of 240 mg/100 cm^3 would legally be able to drive in these regions.

    A   5 hours

    B   10 hours

    C   15 hours

    D   20 hours

[Turn over

16. A number of students were trained to carry out a complex task. Some competed with one another, others worked in isolation.

The graph below shows the number of errors recorded in the training process.

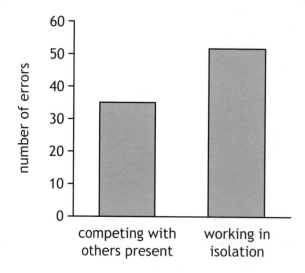

Which process is likely to have caused the difference in the results?

A   Deindividuation

B   Social facilitation

C   Shaping

D   Internalisation

17. The pathogen for the disease tuberculosis (TB) evades the specific immune response by

A   surviving within phagocytes

B   attacking lymphocytes

C   attacking phagocytes

D   antigenic variation.

18. The graph below shows the average growth rate of body organs in males.

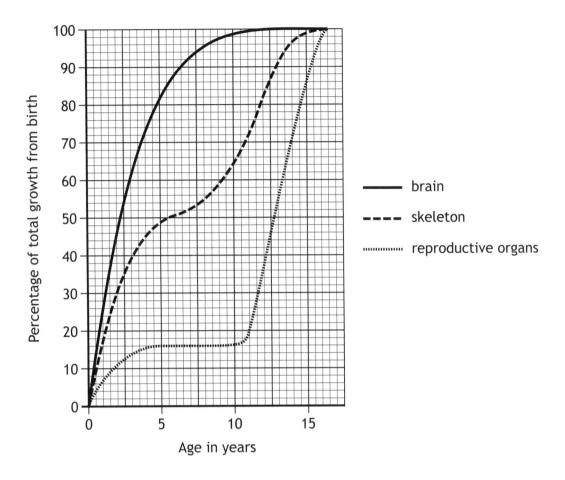

What is the ratio of total growth of brain to skeleton in an 8 year old child?

A   11 : 3

B   3 : 11

C   19 : 11

D   11 : 19

19. Failure in regulation of the immune system leading to an autoimmune disease is caused by a

A   B lymphocyte immune response to self antigens.

B   T lymphocyte immune response to self antigens.

C   B lymphocyte immune response to foreign antigens.

D   T lymphocyte immune response to foreign antigens.

[Turn over for Question 20 on *Page twelve*

20. Blood tests to measure the number of white blood cells (leucocytes) are often used to indicate infection and/or illness.

Leucopenia, due to starvation or malnutrition, is indicated by white blood cell numbers dropping below 4×10^9/litre.

Leucocytosis, due to fever or tissue damage, is indicated by white blood cell numbers temporarily increasing to 11×10^9/litre.

Leukaemia, due to DNA damage and cell division, is indicated by white blood cell numbers permanently increasing.

The following graphs show the white blood cell count of four patients over 20 weeks.

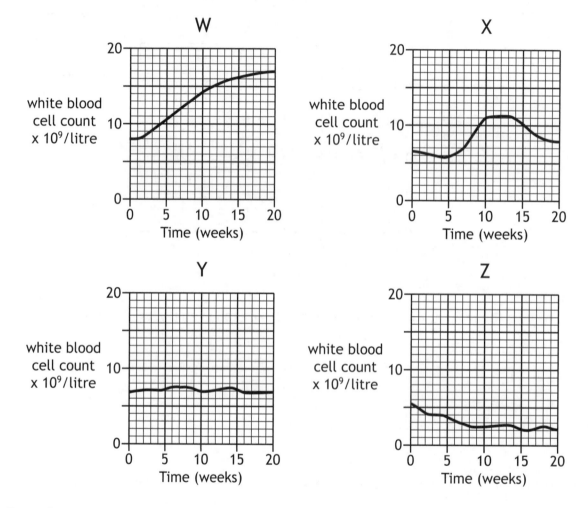

From the graphs, identify the patients.

	Leukaemia	Leucocytosis	Leucopenia
A	Y	X	Z
B	Z	W	Y
C	W	X	Z
D	W	Y	X

[END OF SECTION 1. NOW ATTEMPT THE QUESTIONS IN SECTION 2 OF YOUR QUESTION AND ANSWER BOOKLET.]

H

National
Qualifications
2015

Mark

X740/76/01

**Human Biology
Section 1 — Answer Grid
and Section 2**

WEDNESDAY, 13 MAY

1:00 PM – 3:30 PM

Fill in these boxes and read what is printed below.

Full name of centre

Town

Forename(s)

Surname

Number of seat

Date of birth

Day	Month	Year	Scottish candidate number

Total marks — 100

SECTION 1 — 20 marks

Attempt ALL questions.

Instructions for completion of Section 1 are given on *Page two*.

SECTION 2 — 80 marks

Attempt ALL questions.

Write your answers clearly in the spaces provided in this booklet. Additional space for answers and rough work is provided at the end of this booklet. If you use this space you must clearly identify the question number you are attempting. Any rough work must be written in this booklet. You should score through your rough work when you have written your final copy.

Use **blue** or **black** ink.

Before leaving the examination room you must give this booklet to the Invigilator; if you do not, you may lose all the marks for this paper.

SECTION 1— 20 marks

The questions for Section 1 are contained in the question paper X740/76/02.
Read these and record your answers on the answer grid on *Page three* opposite.
Use **blue** or **black** ink. Do NOT use gel pens or pencil.

1. The answer to each question is **either** A, B, C or D. Decide what your answer is, then fill in the appropriate bubble (see sample question below).

2. There is **only one correct** answer to each question.

3. Any rough working should be done on the additional space for answers and rough work at the end of this booklet.

Sample Question

The digestive enzyme pepsin is most active in the

    A   mouth

    B   stomach

    C   duodenum

    D   pancreas.

The correct answer is **B**—stomach. The answer **B** bubble has been clearly filled in (see below).

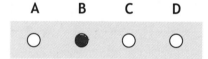

Changing an answer

If you decide to change your answer, cancel your first answer by putting a cross through it (see below) and fill in the answer you want. The answer below has been changed to **D**.

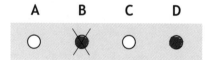

If you then decide to change back to an answer you have already scored out, put a tick (✓) to the **right** of the answer you want, as shown below:

    or

SECTION 1 — Answer Grid

	A	B	C	D
1	○	○	○	○
2	○	○	○	○
3	○	○	○	○
4	○	○	○	○
5	○	○	○	○
6	○	○	○	○
7	○	○	○	○
8	○	○	○	○
9	○	○	○	○
10	○	○	○	○
11	○	○	○	○
12	○	○	○	○
13	○	○	○	○
14	○	○	○	○
15	○	○	○	○
16	○	○	○	○
17	○	○	○	○
18	○	○	○	○
19	○	○	○	○
20	○	○	○	○

[BLANK PAGE]

DO NOT WRITE ON THIS PAGE

Page five

[Turn over for Section 2 on *Page six*

DO NOT WRITE ON THIS PAGE

MARKS | DO NOT WRITE IN THIS MARGIN

SECTION 2 — 80 marks

Attempt ALL questions

Note that Question 14 contains a choice

1. The diagram below represents an embryo in the early stages of development and identifies the inner cell mass which is made up of stem cells.

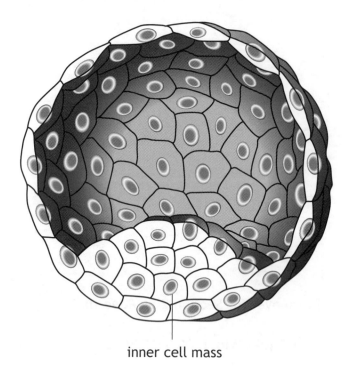

inner cell mass

(a) State one feature of stem cells.

1

_____

_____

(b) Stem cells are also found in tissues throughout the adult body.

Explain how the function of tissue stem cells differs from that of the stem cells found in the inner cell mass of an embryo.

1

_____

_____

MARKS | DO NOT WRITE IN THIS MARGIN

1. (continued)

(c) Stem cells have uses in both therapy and research.

(i) It has been proposed that tissue cells could be used to repair severely damaged muscle tissue.

Suggest how this might be done.   **1**

_____

_____

_____

(ii) State how stem cells can be used as model cells in medical research.   **1**

_____

_____

[Turn over

MARKS | DO NOT WRITE IN THIS MARGIN

2. Glycogen storage disease is an inherited condition in which the enzyme glycogen synthase does not function.

This enzyme normally catalyses one step in the conversion of glucose to glycogen, for storage, as shown in the diagram below.

        enzyme 1   *enzyme 2*   *glycogen synthase*

glucose ⟶ compound A ⟶ compound B ⟶ glycogen

(a) State the term which describes a metabolic pathway in which simple molecules are built up into complex molecules.

1

(b) (i) Describe how the genetic code for glycogen synthase might be altered in an individual with the disease.

1

   (ii) Explain why this altered genetic code fails to produce glycogen synthase.

1

(c) Suggest why individuals with glycogen storage disease might develop abnormally low blood glucose levels during exercise.

1

(d) One form of glycogen storage disease is caused by a gene which is recessive and sex-linked.

Describe a pattern of inheritance, shown by a family history, which would indicate that the condition is

2

recessive _____

_____

sex-linked _____

_____

MARKS

3. Most skin cancers are caused by overexposure to ultraviolet (UV) radiation from the sun or sunbeds. UV radiation damages the DNA in skin cells. Cells normally repair this damage but those which cannot may become cancerous.

A student designed an investigation which used UV-sensitive yeast cells to show the damaging effect of UV radiation. These yeast cells cannot repair DNA damage and die after exposure to UV radiation.

A suspension of UV-sensitive yeast cells was added to dishes which contained a gel that had all the nutrients the yeast needed to grow. The dishes were then exposed to UV radiation for different lengths of time. After exposure, the dishes were placed in an incubator and each of the surviving yeast cells left to grow into a colony on the gel. The number of these colonies was then counted.

The diagram below illustrates this procedure.

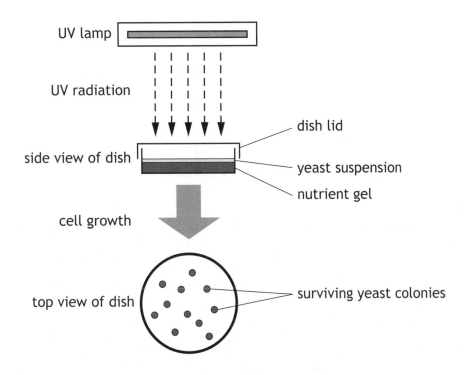

(a) List **two** variables which should be kept constant during this investigation.

2

1 _____

2 _____

[Turn over

3.　(continued)

(b)　The results of the investigation are shown in **Table 1** below.

Table 1 — Yeast growth after exposure to UV radiation

Length of time of exposure (minutes)	Number of yeast colonies growing
10	58
20	32
30	15
40	4
50	1
60	0

(i)　Plot a line graph to illustrate the results of the investigation.　　2

(Additional graph paper, if required can be found on *Page thirty-one*)

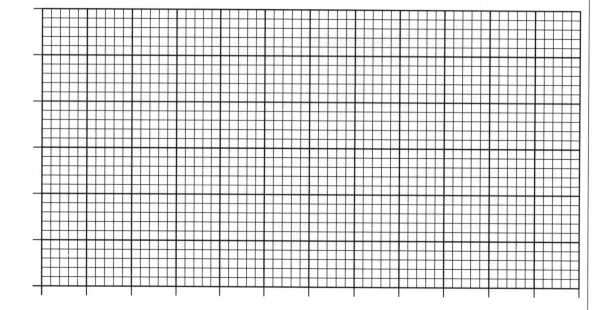

(ii)　State a conclusion that can be drawn from these results.　　1

_____

_____

(iii)　State how the reliability of the results in this investigation could be improved.　　1

_____

_____

MARKS | DO NOT WRITE IN THIS MARGIN

3. **(continued)**

(c) Sunscreens work by blocking UV radiation, preventing it from entering skin cells and causing damage to the DNA, which results in sunburn.

Sunscreens are labelled with a Sun Protection Factor (SPF).

When a sunscreen of SPF 15 is applied to the skin, it will take 15 times longer to burn compared to having no sunscreen applied.

The student carried out a second investigation using UV-sensitive yeast.

The dishes were prepared as before but this time the lids of the dishes were coated with sunscreen of different SPFs. The dishes were then exposed to UV radiation for 30 minutes. After exposure, the dishes were placed in an incubator and the surviving yeast cells left to grow into colonies. The results are shown in **Table 2** below.

Table 2 — Yeast growth after the use of sunscreen protection

Sunscreen used to coat lid (SPF)	Number of yeast colonies growing
6	20
15	72
35	74
50	75

(i) Use the information from **Tables 1 and 2** to calculate the percentage increase in yeast cell survival when a sunscreen of SPF 50 is used to coat the lid.

Space for calculation

_____ %

1

(ii) Official health advice recommends that people should use a sunscreen of SPF 15 when sunbathing for 30 minutes.

State how the results of this investigation support this recommendation.

_____

_____

_____

1

(iii) If skin starts to burn after 10 minutes in strong sunlight, calculate for how long a sunscreen of SPF 35 would protect the skin.

Space for calculation

1

MARKS | DO NOT WRITE IN THIS MARGIN

4. The diagram below represents **three** chemical reactions in the energy investment phase of glycolysis.

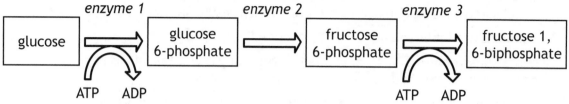

(a) Identify the information, shown **in the diagram**, which confirms that this is the energy investment phase of glycolysis.

1

_____

_____

(b) Enzyme 1 is activated by the binding of magnesium ions.

Suggest how the binding of these ions leads to an increase in enzyme activity.

1

_____

_____

(c) Choose an enzyme shown **in the diagram** which is catalysing a phosphorylation reaction.

Circle **one** enzyme —   Enzyme 1   Enzyme 2   Enzyme 3

Explain what is meant by phosphorylation.

1

_____

_____

(d) The conversion of glucose 6-phosphate to fructose 6-phosphate is a reversible reaction.

Describe the circumstances under which this reaction would go in the opposite direction to that shown in the diagram.

1

_____

_____

(e) Following the energy investment phase, glycolysis enters the energy pay off stage, during which ATP is produced.

Enzyme 3 is phosphofructokinase which is inhibited by a build-up of ATP.

Explain how this feedback mechanism conserves the cell's resources.

1

_____

_____

MARKS | DO NOT WRITE IN THIS MARGIN

5. Muscle cells utilise a variety of energy systems during strenuous activity.

   (a) Creatine phosphate is found in muscle cells.

       (i) Describe how creatine phosphate supports strenuous muscle activity.   **2**

       _____

       _____

       _____

       _____

       (ii) Explain why this support is not provided to strenuous activities beyond the first 10 seconds.   **1**

       _____

       _____

   (b) Name the substance which builds up in muscle cells as they become fatigued.   **1**

       _____

       _____

   (c) Choose a sporting activity and decide whether slow twitch or fast twitch muscle fibres would be best suited for the activity.

       Sporting Activity _____

       Slow twitch [ ]   Fast twitch [ ]

       Give reasons to justify your choice of muscle fibre.   **3**

       _____

       _____

       _____

       _____

       _____

       _____

       _____

       _____

MARKS | DO NOT WRITE IN THIS MARGIN

6. Chorionic villus sampling (CVS) is a technique which can be used during antenatal screening. The cells obtained from CVS are used to prepare a karyotype.

(a) The diagram below shows the uterus of a pregnant woman with a section of the placenta enlarged.

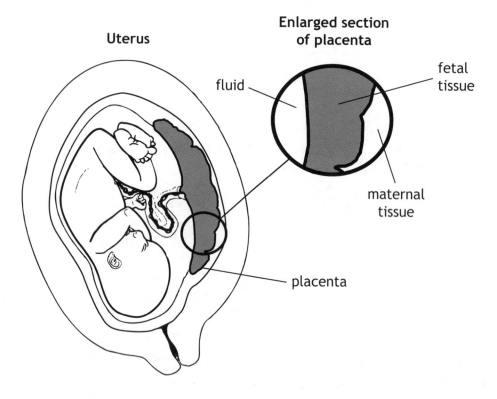

Uterus

Enlarged section of placenta

fluid

fetal tissue

maternal tissue

placenta

(i) Place a cross (X) on the diagram of the **enlarged section of placenta** to indicate the area from which cells are removed during CVS.   1

(ii) Describe the process by which a karyotype is produced from cells removed during CVS.   2

_____

_____

_____

_____

(iii) Suggest an advantage of using CVS rather than amniocentesis during antenatal screening.   1

_____

_____

MARKS | DO NOT WRITE IN THIS MARGIN

6. (continued)

(b) Name the type of antenatal screening tests which are routinely carried out to monitor the concentration of certain substances, such as protein, in a pregnant woman's blood.   1

_____

[Turn over

7. The graph below contains information about the body mass index (BMI) of Scottish children in 2009.

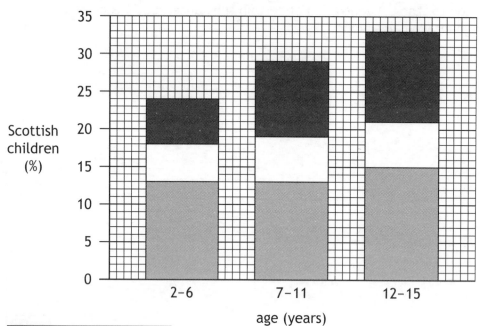

(a) State the percentage of children aged 12 to 15 who had a BMI of more than 30 in 2009.

Space for calculation

_____ %

(b) Suggest reasons why the percentage of obese children increased between the ages of 2 and 15.

_____

_____

MARKS | DO NOT WRITE IN THIS MARGIN

7. (continued)

(c) Explain how BMI is calculated.   1

_____

_____

(d) Suggest how children could be encouraged to maintain a healthy BMI by use of the following processes.   2

Identification _____

_____

Internalisation _____

_____

[Turn over

8. The heart rate and stroke volume of a 40 year old cyclist were monitored as he used an exercise bike.

The cyclist was told to pedal at a constant rate as his work level was gradually raised by increasing the resistance to pedalling.

The graph below shows the changes that occurred in the cyclist's heart rate and stroke volume at seven different work levels.

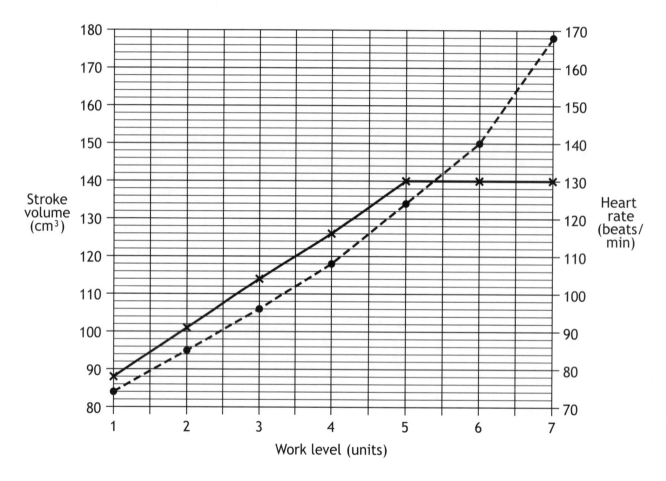

KEY

×———————× stroke volume - volume of blood pumped out per heartbeat

●－－－－● heart rate - beats of heart per minute

MARKS | DO NOT WRITE IN THIS MARGIN

8. (continued)

(a) Use **data from the graph** to describe the changes that occurred in the cyclist's stroke volume when the work level increased from 1 to 7 units.   2

_____

_____

_____

_____

(b) State what the cyclist's heart rate was when his stroke volume was 120 cm^3.   1

Space for calculation

_____ beats/min

(c) Cardiac output is the volume of blood leaving the heart in one minute. It is calculated using the formula shown below.

cardiac output = heart rate × stroke volume

Calculate the cyclist's cardiac output when his work level was 6 units.   1

Space for calculation

_____ cm^3/min

[Turn over

MARKS | DO NOT WRITE IN THIS MARGIN

8. **(continued)**

(d) The table below shows the recommended minimum heart rates that cyclists of different ages should maintain in order to either metabolise fat or improve their fitness.

Age	Minimum heart rate for metabolising fat (beats/min)	Minimum heart rate for improving fitness (beats/min)
10	136	168
20	130	160
30	123	152
40	116	144
50	110	136
60	104	128

(i) Use information from the **table** and the **graph** to determine the work level that the cyclist should maintain in order to metabolise fat.

1

_____ units

(ii) Use information from the **table** to predict the minimum heart rate for improving the fitness of a 70 year old.

1

_____

(iii) As an individual gets older, their minimum heart rate for improving fitness decreases.

Use the information from the **table** to calculate the percentage decrease that occurs between the ages of 10 and 60 years.

1

Space for calculation

_____ %

MARKS | DO NOT WRITE IN THIS MARGIN

9. The diagram below shows some nerve cells involved in a neural reward pathway.

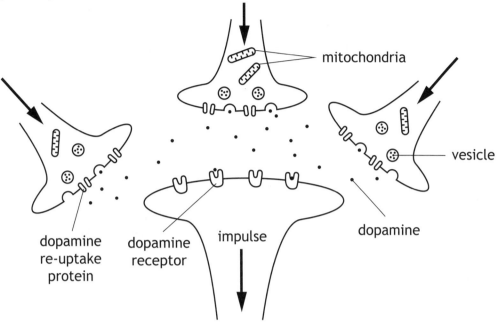

(a) Using information from the diagram, explain what is meant by the term "summation".   1

_____

_____

(b) Suggest a function for the mitochondria shown in the diagram.   1

_____

_____

(c) Cocaine is a recreational drug that has an effect at this synapse.

Cocaine binds to the dopamine re-uptake proteins. As a result, the reward pathway is stimulated for longer.

Suggest how cocaine produces this effect.   2

_____

_____

_____

_____

[Turn over

MARKS | DO NOT WRITE IN THIS MARGIN

10. A biology student produced the following diagram as a memory aid to help her learn about transport in plants.

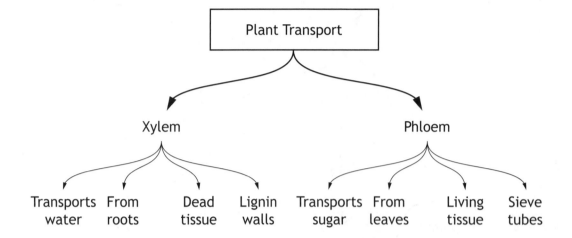

(a) In producing this diagram the student used various methods to learn the information.

Other than rehearsal, name **two** methods that she used and describe how they helped her transfer the information into her long-term memory.   2

1   Method _____

Description _____

_____

2   Method _____

Description _____

_____

(b) Any information which is not transferred into long-term memory is displaced. Explain why displacement occurs.   1

_____

_____

(c) The student is storing a record of facts as she learns this information. State the part of the brain in which such memories are stored.   1

_____

MARKS | DO NOT WRITE IN THIS MARGIN

11. Various types of white blood cell are involved in the non-specific immune response.

(a) Describe the role of each of the following cells in the non-specific defence of the body.

   (i) Mast cells _____   **2**

_____

_____

_____

   (ii) Natural killer (NK) cells _____   **1**

_____

_____

_____

(b) Explain how the presence of phagocytes is important in the activation of T lymphocytes.   **2**

_____

_____

_____

_____

[Turn over

12. HIV is a virus which invades the cells of the immune system.

People infected with HIV may not show symptoms for many years.

AIDS is the condition which may develop from HIV infection, resulting in death.

The graph below shows the number of people in the world infected with HIV, from 1990 to 2010. It also shows the number of people who died from AIDS during this period.

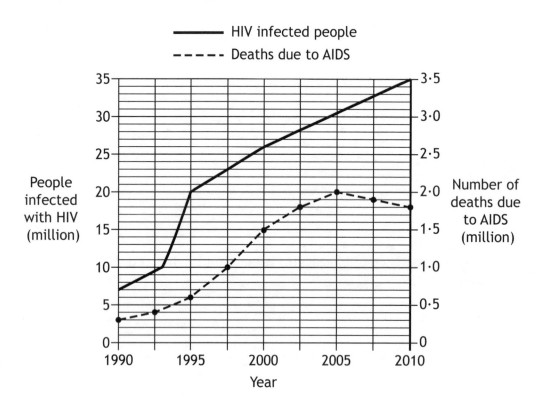

(a) State how many people were infected with HIV in the year 2000.   1

_____

(b) State how many people died from AIDS when 20 million people in the world were infected with HIV.   1

_____

(c) Calculate the percentage of HIV-infected people who died from AIDS in 2010.   1

Space for calculation

_____ %

MARKS | DO NOT WRITE IN THIS MARGIN

12. (continued)

(d) Describe the evidence from the graph which suggests that the rate of people becoming infected with HIV was greatest between 1993 and 1995.

1

[Turn over

MARKS | DO NOT WRITE IN THIS MARGIN

13. A scientist investigated the effectiveness of four different types of influenza vaccine. A total of 2000 volunteers from a Scottish community were divided into four groups.

Each group was injected with a different vaccine.

The number who developed influenza during the following years was recorded.

The results are shown in the table below.

Type of influenza vaccine	Developed influenza	Did not develop influenza	Total
P	35	495	530
Q	25	455	480
R	24	496	520
S	17		

(a) (i) Suggest **one** way in which the scientist could minimise variation between the four groups of volunteers.   1

_____

_____

(ii) **Complete the table** for the volunteers who received type S vaccine.   1

(iii) State which of the vaccines P, Q or R was most effective in this investigation.   1

_____

(b) Explain why vaccines usually contain an adjuvant.   1

_____

_____

(c) In 1918 fifty million people died in a global outbreak of influenza.

State the term used to describe such an outbreak.   1

_____

MARKS | DO NOT WRITE IN THIS MARGIN

14. Answer **either** A **or** B in the space below.

Labelled diagrams may be used where appropriate.

A   Describe hormonal control of the menstrual cycle under the following headings:

    (i)   events leading to ovulation;   **6**

    (ii) events following ovulation.   **4**

OR

B   Describe the cardiac cycle under the following headings:

    (i)   the conducting system of the heart;   **5**

    (ii) nervous control of the cardiac cycle.   **5**

[Turn over

ADDITIONAL SPACE FOR ANSWER TO QUESTION 14

MARKS | DO NOT WRITE IN THIS MARGIN

[END OF QUESTION PAPER]

MARKS DO NOT WRITE IN THIS MARGIN

ADDITIONAL SPACE FOR ANSWERS AND ROUGH WORK

ADDITIONAL SPACE FOR ANSWERS AND ROUGH WORK

MARKS | DO NOT WRITE IN THIS MARGIN

MARKS DO NOT WRITE IN THIS MARGIN

ADDITIONAL GRAPH PAPER FOR QUESTION 3 (b) (i)

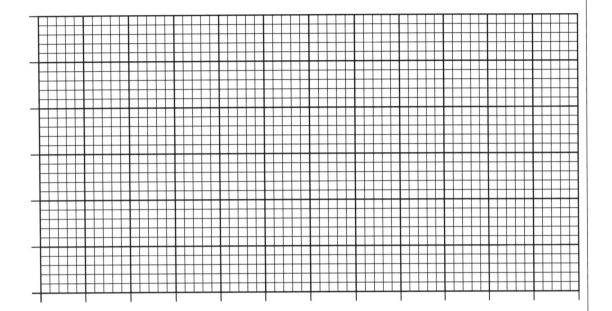

[BLANK PAGE]

DO NOT WRITE ON THIS PAGE

HIGHER

2016

X740/76/02

Human Biology
Section 1 — Questions

MONDAY, 9 MAY

1:00 PM – 3:30 PM

Instructions for the completion of Section 1 are given on *Page two* of your question and answer booklet X740/76/01.

Record your answers on the answer grid on *Page three* of your question and answer booklet.

Before leaving the examination room you must give your question and answer booklet to the Invigilator; if you do not, you may lose all the marks for this paper.

SECTION 1 — 20 marks

Attempt ALL questions

1. In a developing embryo, tissues such as muscle and nerve are produced by

   A   somatic cells dividing by meiosis

   B   germline cells dividing by meiosis

   C   somatic cells dividing by mitosis

   D   germline cells dividing by mitosis.

2. A genetic disorder of the nervous system results from a mutation in which a nucleotide is inserted into a gene.

   Which of the following types of mutation causes this genetic disorder?

   A   nonsense

   B   missense

   C   translocation

   D   frame-shift

3. The following steps occur during the Polymerase Chain Reaction (PCR).

   1.   Binding of primer

   2.   Replication of DNA

   3.   Heating of sample DNA

   4.   Separation of DNA strands

   In which sequence do these steps occur?

   A $1 \rightarrow 2 \rightarrow 4 \rightarrow 3$

   B $1 \rightarrow 2 \rightarrow 3 \rightarrow 4$

   C $3 \rightarrow 4 \rightarrow 1 \rightarrow 2$

   D $3 \rightarrow 4 \rightarrow 2 \rightarrow 1$

4. The diagrams below represent the shapes of an enzyme molecule and its substrate.

Enzyme molecule   *Substrate molecule*

Which row in the table below shows the possible shapes of two types of molecule that could inhibit the enzyme above?

	Competitive Inhibitor	Non-competitive Inhibitor
A	⬤	◣◠
B	▲	▢
C	▢	⬤
D	◢◣	▢

5. During glycolysis, dehydrogenase enzymes catalyse the

A   removal of hydrogen ions from $NADH_2$

B   removal of hydrogen ions from citrate

C   transfer of hydrogen ions to glucose

D   transfer of hydrogen ions to NAD.

[Turn over

Page three

6. The diagram below represents a mitochondrion which has been magnified 10 000 times.

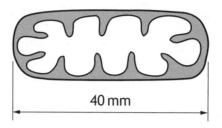

40 mm

What is the actual length of this mitochondrion?

(1 mm = 1000 micrometres)

A   0·04 micrometres

B   0·4 micrometres

C   4 micrometres

D   40 micrometres

7. The diagram below represents some of the processes which occur at the inner membrane of a mitochondrion.

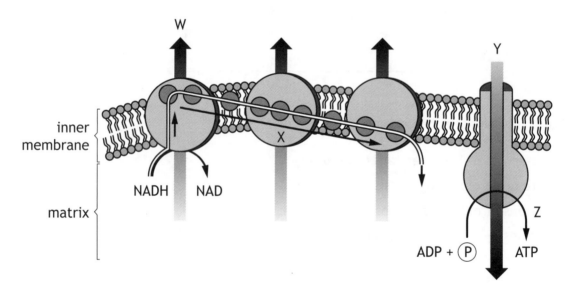

Which letter represents the transfer of high energy electrons?

A   W

B   X

C   Y

D   Z

8. During cellular respiration, the activity of phosphofructokinase can be inhibited by

    A   ATP and citrate

    B   ADP and citrate

    C   ATP and lactic acid

    D   ADP and lactic acid.

9. The graph below shows changes which occur in the masses of protein, fat and carbohydrate in a person's body during seven weeks without food.

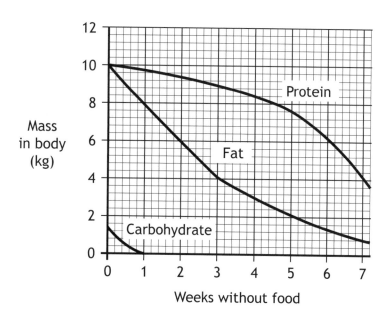

The person's starting mass was 60 kg.

Predict their mass after two weeks without food.

    A   57 kg

    B   54 kg

    C   50 kg

    D   43 kg

[Turn over

10. The diagram below represents connections between parts of the male reproductive system. Which arrow in the diagram does **not** represent a male reproductive hormone?

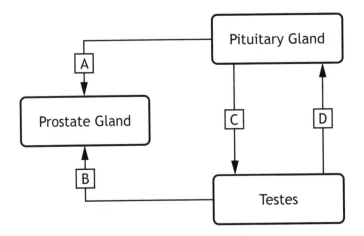

11. The graph below shows the chance of a woman becoming pregnant, following sexual intercourse, on the days before and after ovulation.

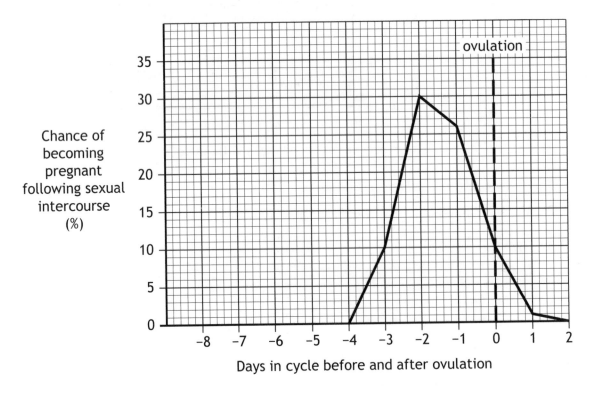

Days in cycle before and after ovulation

This woman has a 28 day menstrual cycle and ovulates on the 3rd of May.

On which day in May would having sexual intercourse give her the best chance of becoming pregnant?

A   3rd May

B   17th May

C   29th May

D   31st May

12. In the treatment of infertility, ovulation can be stimulated by drugs that prevent the negative feedback effect of

   A   oestrogen on LH secretion

   B   oestrogen on FSH secretion

   C   progesterone on LH secretion

   D   progesterone on FSH secretion.

13. During antenatal care, which **two** techniques can be used to obtain cells for production of a karyotype?

   A   Chorionic villus sampling (CVS) and amniocentesis

   B   Ultrasound imaging and chorionic villus sampling (CVS)

   C   Amniocentesis and pre-implantation genetic diagnosis (PGD)

   D   Pre-implantation genetic diagnosis (PGD) and ultrasound imaging

14. The inheritance of an allele for deafness is shown in the family tree below.

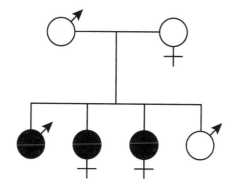

Key	
♀	Unaffected female
♂	Unaffected male
●	Affected female
●	Affected male

This condition is controlled by an allele which is

   A   dominant and sex-linked

   B   recessive and sex-linked

   C   dominant and not sex-linked

   D   recessive and not sex-linked.

[Turn over

15. Which of the following memories would be stored in the limbic system only?

A   The tune to your favourite song.

B   How to keep three balls in the air when juggling.

C   The route to your bed across your bedroom in the dark.

D   The taste of your favourite food.

16. Playing cards normally have hearts and diamonds in red, and spades and clubs in black.

An investigation showed that the speed and accuracy in recognising the cards decreased when the colours were reversed, for example when hearts appeared black.

This result was most likely to have been caused by the effect of

A   a perceptual set

B   a binocular disparity

C   a segregation into figure and ground

D   an organisation into coherent patterns.

17. The diagram below shows the ages at which infants are able to walk unaided.

    The left end of the bar shows the age at which 25% of infants can walk unaided.

    The right end of the bar shows the age at which 90% of infants can walk unaided.

    The vertical line on the bar shows the age at which 50% of infants can walk unaided.

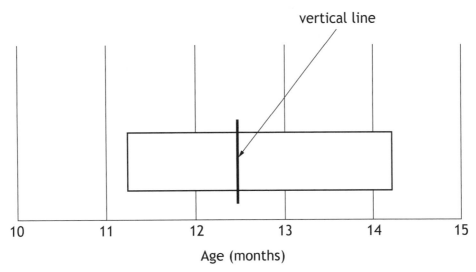

Age (months)

If 24 infants, aged 12 months, were tested, how many would be expected to walk unaided?

A   6

B   10

C   14

D   18

18. The table below contains information about two groups of students who were asked to construct a paper model from a set of instructions.

Group	Arrangement of students	Average time to complete model (s)
1	all students together in one room	105
2	each student in a separate room	140

The improved performance of the students in group 1 is likely to be due to

A   shaping

B   discrimination

C   deindividuation

D   social facilitation.

19. When tissue is damaged, mast cells release histamine which **immediately** results in

    A   an accumulation of phagocytes

    B   increased delivery of antimicrobial proteins and clotting elements

    C   increased localised blood vessel dilation and capillary permeability

    D   stimulation of a specific immune response by activating lymphocytes.

20. In Scotland cases of influenza are always present but occasionally they rise to unusually high levels. In this case, the disease is said to have changed from being

    A   epidemic to endemic

    B   endemic to epidemic

    C   sporadic to epidemic

    D   endemic to sporadic.

**[END OF SECTION 1. NOW ATTEMPT THE QUESTIONS IN SECTION 2
OF YOUR QUESTION AND ANSWER BOOKLET.]**

FOR OFFICIAL USE

National
Qualifications
2016

Mark

X740/76/01

**Human Biology
Section 1 — Answer Grid
and Section 2**

MONDAY, 9 MAY

1:00 PM – 3:30 PM

Fill in these boxes and read what is printed below.

Full name of centre

Town

Forename(s)

Surname

Number of seat

Date of birth

Day	Month	Year	Scottish candidate number

Total marks — 100

SECTION 1 — 20 marks

Attempt ALL questions.

Instructions for the completion of Section 1 are given on *Page two*.

SECTION 2 — 80 marks

Attempt ALL questions.

Write your answers clearly in the spaces provided in this booklet. Additional space for answers and rough work is provided at the end of this booklet. If you use this space you must clearly identify the question number you are attempting. Any rough work must be written in this booklet. You should score through your rough work when you have written your final copy.

Use **blue** or **black** ink.

Before leaving the examination room you must give this booklet to the Invigilator; if you do not, you may lose all the marks for this paper.

SECTION 1 — 20 marks

The questions for Section 1 are contained in the question paper X740/76/02.

Read these and record your answers on the answer grid on *Page three* opposite.

Use **blue** or **black** ink. Do NOT use gel pens or pencil.

1. The answer to each question is **either** A, B, C or D. Decide what your answer is, then fill in the appropriate bubble (see sample question below).

2. There is **only one correct** answer to each question.

3. Any rough working should be done on the additional space for answers and rough work at the end of this booklet.

Sample Question

The digestive enzyme pepsin is most active in the

    A   mouth

    B   stomach

    C   duodenum

    D   pancreas.

The correct answer is **B** — stomach. The answer **B** bubble has been clearly filled in (see below).

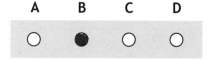

Changing an answer

If you decide to change your answer, cancel your first answer by putting a cross through it (see below) and fill in the answer you want. The answer below has been changed to **D**.

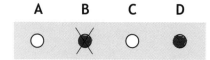

If you then decide to change back to an answer you have already scored out, put a tick (✓) to the **right** of the answer you want, as shown below:

SECTION 1 — Answer Grid

	A	B	C	D
1	○	○	○	○
2	○	○	○	○
3	○	○	○	○
4	○	○	○	○
5	○	○	○	○
6	○	○	○	○
7	○	○	○	○
8	○	○	○	○
9	○	○	○	○
10	○	○	○	○
11	○	○	○	○
12	○	○	○	○
13	○	○	○	○
14	○	○	○	○
15	○	○	○	○
16	○	○	○	○
17	○	○	○	○
18	○	○	○	○
19	○	○	○	○
20	○	○	○	○

[BLANK PAGE]

DO NOT WRITE ON THIS PAGE

MARKS

DO NOT
WRITE IN
THIS
MARGIN

SECTION 2 – 80 marks

Attempt ALL questions

Note that Question 13 contains a choice.

1. The diagram below represents a stage in the process of DNA replication.

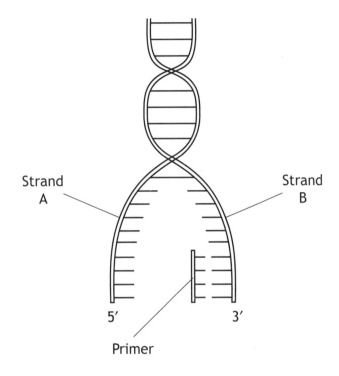

(a) (i) Name the type of bond which links the primer to strand B.   1

_____

(ii) Name the chemical group found at the 5′ end of a DNA strand.   1

_____

(b) Strand B is replicated continuously while strand A can only be replicated in fragments.

Explain why the strands are replicated in different ways.   1

_____

_____

_____

MARKS | DO NOT WRITE IN THIS MARGIN

1. **(continued)**

(c) Describe the role of the following enzymes in DNA replication.   2

DNA polymerase _____

_____

Ligase _____

_____

1. **(continued)**

MARKS

DO NOT
WRITE IN
THIS
MARGIN

2. At the start of polypeptide synthesis in a cell, DNA transcription occurs in the nucleus to form mRNA.

The sequence of bases from a section of a DNA strand is shown below.

…... C A C G A T C G A T A G G A T …...

(a) (i) State the sequence of bases in the primary mRNA transcript formed from this strand of DNA.

1

_____

(ii) State the term used to describe the coding regions of a primary mRNA transcript.

1

_____

(iii) Name the process by which the coding regions of a primary mRNA transcript are joined together to produce a mature mRNA transcript.

1

_____

(iv) The sequence of bases in the mature mRNA transcript, formed from the section of the DNA strand, is shown below.

…... G U G C U A U C C U A …...

Using this mature mRNA transcript, state the order of bases in the intron present in the primary mRNA transcript.

1

_____

(b) State the location for the translation of a mature mRNA transcript into a polypeptide.

1

_____

(c) Describe **one** form of post-translational modification of a polypeptide.

1

_____

_____

[Turn over

MARKS | DO NOT WRITE IN THIS MARGIN

3. A naturally occurring cell protein (nm23) has been shown to inhibit the activity of cancer cells.

Individuals produce varying levels of this protein depending on their genetic make-up.

The graph below shows the results of a 9 year study of women diagnosed with breast cancer. The women were divided into two groups according to their production of the protein.

Key: ——— patients with normal levels of the protein
     •••• patients with low levels of the protein

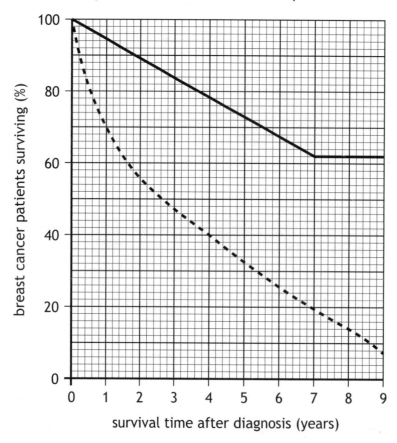

(a) (i) In a city, 1000 women were diagnosed with breast cancer.

Of these women, 900 had normal levels of the protein while 100 had low levels.

Using the results from the study, calculate how many of the 1000 women would be expected to survive for 4 years after diagnosis.   **1**

Space for calculation

MARKS

3. (a) (continued)

(ii) **Use data from the graph** to describe the changes in the percentage of surviving breast cancer patients with normal levels of the protein during the study.   2

_____

_____

_____

_____

_____

(b) Describe how cancer can develop and spread through the body.   3

_____

_____

_____

_____

_____

_____

_____

_____

_____

[Turn over

MARKS | DO NOT WRITE IN THIS MARGIN

4. A student carried out an investigation into the effect of physical activity on respiration rate.

The rate of respiration of six individuals was measured after carrying out three different activities for five minutes.

Immediately after completing the activity, each individual breathed into a bottle containing a pH indicator solution. This indicator changes colour from blue to yellow in the presence of a high concentration of carbon dioxide.

Figure 1 – Apparatus used

Table 1 – Results of Investigation

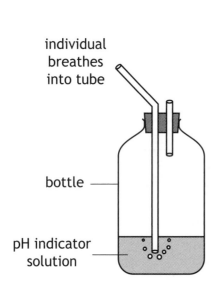

individual breathes into tube

bottle

pH indicator solution

Individual	Time taken for pH indicator to turn yellow (s)		
	Activity 1 resting	Activity 2 walking	Activity 3 running
1	33	28	22
2	29	14	11
3	22	16	12
4	30	26	20
5	44	35	22
6		31	21
Average time taken (s)	33	25	18

(a) State **two** variables which would have to be kept constant when setting up the apparatus shown in **Figure 1**.   2

1 _____

2 _____

(b) Calculate the time taken for the indicator to turn yellow after individual 6 had completed Activity 1.   1

Space for calculation

_____ s

MARKS

DO NOT WRITE IN THIS MARGIN

4. (continued)

(c) Describe how the student increased the reliability of the results.   **1**

_____

_____

(d) Construct a bar graph to show the average results obtained in this investigation.   **2**

(Additional graph paper, if required, can be found on _Page thirty-one_)

(e) State a conclusion that can be drawn from the results of this investigation.   **1**

_____

_____

(f) Suggest an explanation for the results obtained in this investigation.   **1**

_____

_____

_____

[Turn over

MARKS | DO NOT WRITE IN THIS MARGIN

5. The diagram below represents the structure of the heart and its associated blood vessels.

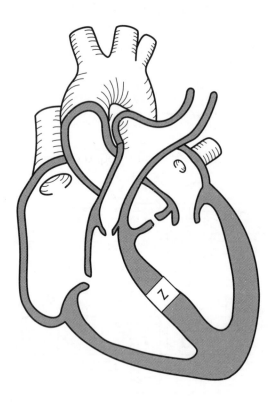

(a) On the diagram, **label** the pulmonary artery with the letter P.   1

(b) Sometimes babies can be born with a ventricular septal defect (VSD) in which a "hole" occurs at point Z in the heart.

Explain how the presence of this hole would affect the oxygen concentration of the blood leaving the heart through the aorta.   2

_____

_____

_____

_____

MARKS

5. (continued)

(c) Babies with a VSD sometimes have irregular heart rhythms. This can be detected by recording the electrical activity from the heart.

(i) Name the chamber of the heart in which this electrical activity originates.

1

_____

(ii) Name the type of graph that displays such patterns of electrical activity.

1

_____

(d) Babies with a VSD often have a lower stroke volume than babies who have a normal heart structure.

Despite this, both groups of babies often have similar cardiac outputs.

Suggest how babies with a VSD are able to achieve a similar cardiac output to babies with a normal heart structure.

1

_____

_____

[Turn over

6. The graph below shows the number of stroke patients in different groups from Scotland and England in 2007.

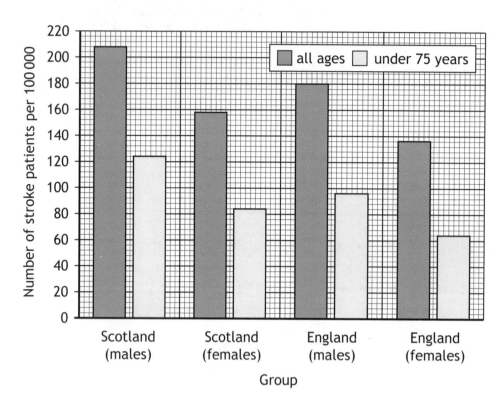

(a) (i) Calculate the difference in the number of male stroke patients of all ages in Scotland and England in 2007.

Space for calculation

1

_____ per 100 000

(ii) Explain the importance of presenting the data as the number of stroke patients **per 100 000**.

1

_____

_____

MARKS | DO NOT WRITE IN THIS MARGIN

6. (a) (continued)

   (iii) Scotland's population was 5·1 million in 2007.

   Calculate the number of female stroke patients in Scotland under 75 years of age in this year.

   Space for calculation

   1

   _____

   (iv) Express, as a simple whole number ratio, the number of male stroke patients under 75 years of age compared to female stroke patients under 75 years of age in England in 2007.

   Space for calculation

   1

   _____ : _____
   male patients   female patients

(b) Describe what causes a stroke.

   1

   _____

   _____

(c) Paralysis occurs when voluntary muscle is unable to contract.

   Explain how a stroke could lead to muscle paralysis on the left side of the body.

   2

   _____

   _____

   _____

   _____

   _____

7. The table below contains information about five obese patients who attended a weight loss clinic for 12 weeks.

Patient	Height (m)	Starting weight (kg)	Starting BMI	Final Weight (kg)	Final BMI
P	1·74	92	30·5	82	27·2
Q	1·68	98	34·8	90	32·1
R	1·81	104	31·8	97	29·7
S	1·89	121	33·9	113	31·4
T	1·90	100	32·3	94	

(a) (i) Calculate the final BMI of patient T.   1

Space for calculation

Final BMI = _____

   (ii) State why patient Q was still classed as obese after 12 weeks.   1

_____

_____

(b) Explain why all the patients were advised to exercise regularly to increase their weight loss.   1

_____

_____

_____

MARKS | DO NOT WRITE IN THIS MARGIN

7. **(continued)**

(c) Rugby players may have a BMI which indicates that they are obese.

Suggest why a BMI reading may not be a reliable indicator of obesity in rugby players.

1

_____

_____

_____

[Turn over

1

MARKS | DO NOT WRITE IN THIS MARGIN

8. Statins are drugs which reduce the production of cholesterol in the liver. A year-long trial was carried out to investigate the effects of taking a newly-developed statin on blood cholesterol levels.

Sixty individuals with raised blood cholesterol levels were selected and divided into two groups of thirty.

Individuals in Group 1 were prescribed a capsule, containing 20 mg of the statin, to take each day. Individuals in Group 2 were the control group. At two-monthly intervals, blood samples were taken from all individuals and their blood cholesterol levels measured.

The results are shown in the table below.

Month of trial	Average blood cholesterol level (mmol/l)	
	Group 1	Group 2
0	6·3	6·3
2	6·3	6·3
4	6·3	6·1
6	6·3	6·3
8	5·6	6·1
10	5·3	6·2
12	5·1	6·1

(a) **Using the results in the table**, give **one** reason why this drug might be   2

recommended _____

_____

not recommended _____

_____

(b) Suggest what was prescribed to the individuals in Group 2 during this trial.   1

_____

_____

MARKS | DO NOT WRITE IN THIS MARGIN

8. **(continued)**

(c) Describe the design features which would have been used to ensure that this was both a randomised and a double-blind trial.

2

randomised _____

_____

double-blind _____

_____

(d) The bar graph below summarises the data collected in the final month of this trial.

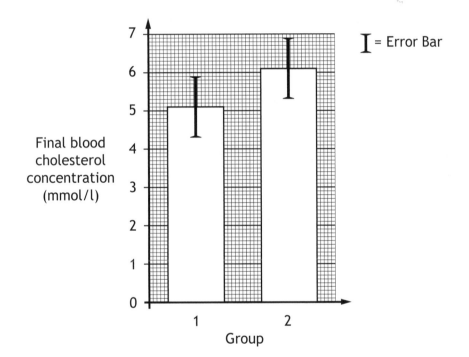

Using evidence from the bar graph, suggest why it was decided that this statin was not worth further development.

1

_____

_____

(e) Describe **one** function of cholesterol in the human body.

1

_____

_____

[Turn over

MARKS | DO NOT WRITE IN THIS MARGIN

9. The diagram below represents areas of high activity in a part of the brain of an individual as a task is described to them which they then complete.

☐ areas of high activity during the description of the task

░ areas of high activity during the completion of the task

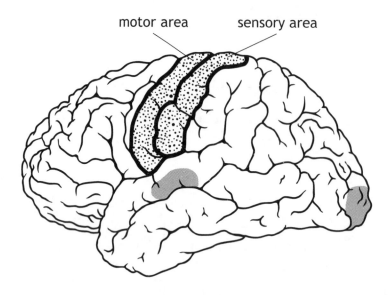

motor area   sensory area

(a) Name the part of the brain shown in the diagram.   1

_____

(b) Explain how the diagram supports the suggestion that there is localisation of function in the brain.   1

_____

_____

(c) Explain the high level of brain activity during the description of the task.   1

_____

_____

MARKS | DO NOT WRITE IN THIS MARGIN

9. **(continued)**

(d) The task was to fold a piece of paper.

Explain why the diagram shows high levels of activity in the sensory and motor areas.

2

Sensory area _____

_____

Motor area _____

_____

[Turn over

MARKS | DO NOT WRITE IN THIS MARGIN

10.   The diagram below shows a synapse in skeletal muscle of a weightlifter.

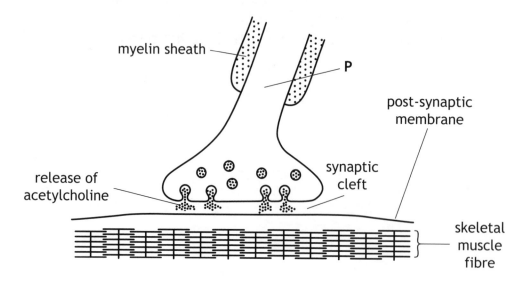

(a)   Name the part of the motor neuron labelled **P**.

1

_____

(b)   Describe what acetylcholine does when it reaches the post-synaptic membrane.

1

_____

_____

(c)   Name the type of skeletal muscle fibres which will be the most common in the arm muscles of a champion weightlifter.

1

_____

MARKS | DO NOT WRITE IN THIS MARGIN

10. **(continued)**

(d) Nicotine is a drug that is an agonist of acetylcholine.

(i) Explain how an agonist works.   1

_____

_____

(ii) Suggest how nicotine induces feelings of pleasure and so reinforces smoking behaviour.   1

_____

_____

[Turn over

MARKS | DO NOT WRITE IN THIS MARGIN

11. The graph below shows the number of cases of measles that occurred in the world between 1980 and 2010. It also shows the global vaccination rate against measles over the same period.

Key: ●————————● number of cases   ●– – – – – – –● vaccination rate

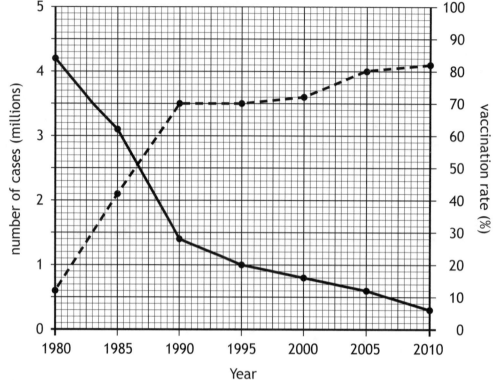

(a) State how many cases of measles there were in 1985.

1

_____

(b) State the vaccination rate when there were 3·5 million cases of measles in the world.

1

_____%

(c) Calculate the percentage decrease in the number of cases of measles between 1995 and 2010.

1

Space for calculation

_____

MARKS | DO NOT WRITE IN THIS MARGIN

11.   (continued)

(d)   In many countries herd immunity has been established against measles.

(i)   Explain why people in these countries who have not been vaccinated are still protected against measles.

1

_____

_____

(ii)   Suggest **one** reason why widespread vaccination programmes against measles are not possible in all countries of the world.

1

_____

_____

(e)   In 2010 the population of the world was 6900 million.

Using the information from the graph, calculate how many people in the world had not been vaccinated against measles in 2010.

1

Space for calculation

_____ million

(f)   The World Health Organisation (WHO) has set a goal of eliminating measles worldwide by 2020.

Explain how the information in the graph indicates that this goal can be achieved.

1

_____

_____

_____

_____

[Turn over

MARKS | DO NOT WRITE IN THIS MARGIN

12. Specific cellular defences are part of the body's immune system and give protection against individual types of pathogen.

(a) (i) Explain how a clonal population of lymphocytes would be formed when a pathogen invades the body.

2

_____

_____

_____

_____

(ii) Describe the role of phagocytes in the specific immune response.

2

_____

_____

_____

_____

(b) Failure of the immune system can lead to conditions such as allergy and autoimmune disease.

Choose **one** of these conditions and complete the table below with information about it.

2

Condition _____

Type of white blood cell involved	Description of immune system failure

MARKS | DO NOT WRITE IN THIS MARGIN

13. Answer **either** A **or** B in the space below.

Labelled diagrams may be used where appropriate.

A   Discuss the causes, development and associated health problems of atherosclerosis.   **8**

OR

B   Discuss the diagnosis, treatment and role of insulin in Type I and Type 2 diabetes.   **8**

MARKS | DO NOT WRITE IN THIS MARGIN

ADDITIONAL SPACE FOR ANSWER to Question 13

[END OF QUESTION PAPER]

ADDITIONAL SPACE FOR ANSWERS AND ROUGH WORK

ADDITIONAL SPACE FOR ANSWERS AND ROUGH WORK

ADDITIONAL SPACE FOR ANSWERS AND ROUGH WORK

Additional graph paper for Question 4 (d)

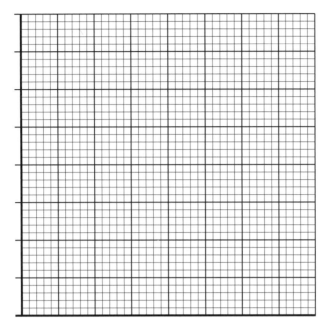

[BLANK PAGE]

DO NOT WRITE ON THIS PAGE

HIGHER

Answers

SQA HIGHER
HUMAN BIOLOGY 2016

HIGHER HUMAN BIOLOGY
2014 SPECIMEN QUESTION PAPER

Section 1

Question	Answer	Mark
1.	A	1
2.	C	1
3.	B	1
4.	D	1
5.	A	1
6.	C	1
7.	B	1
8.	D	1
9.	B	1
10.	A	1
11.	D	1
12.	C	1
13.	C	1
14.	B	1
15.	D	1
16.	A	1
17.	C	1
18.	D	1
19.	B	1
20.	B	1

Section 2

Question			Expected answer(s)	Max mark
1.	(a)		Process — Differentiation. Explanation — only the genes characteristic for that cell are expressed	2
	(b)	(i)	Meiosis.	1
		(ii)	Mutations in germline cells can be passed to offspring (while mutations in somatic cells cannot)	1
	(c)		It is safer than using the drug directly on humans/trial subjects OR Is it right to use embryos to extract stem cells? OR Is it right to deprive sufferers of a potential treatment? OR Is it right to use stem cells rather than animals?	1
2.	(a)		RNA polymerase	1
	(b)		Translation **and** ribosome	1

Question			Expected answer(s)	Max mark
	(c)	(i)	Only one gene is transcribed/forms mRNA OR The primary mRNA only codes for one protein	1
		(ii)	Introns/non-coding regions of genes are removed (in RNA splicing) OR The mature mRNA transcript only contains exons/coding regions of genes	1
3.	(a)		Correct scales and labels on axes Points correctly plotted and line drawn	2
	(b)	(i)	volume of urea solution OR volume of urease solution OR concentration of urease solution OR volume/length of agar/diameter of test tube OR volume/concentration of indicator in agar	1
		(ii)	Temperature of the tube contents/of the test tubes	1
	(c)		The experiment was repeated at each urea concentration (and an average calculated)	1
	(d)		To allow time for the ammonia to (fully) diffuse/spread through the agar/jelly	1
	(e)		As the urea concentration increased more ammonia was produced OR As the urea concentration decreased less ammonia was produced	1
	(f)		48	1
	(g)	(i)	Thiourea blocked the active site on the urease/enzyme	1
		(ii)	Not all active sites were blocked OR some active sites were still available	1
4.	(a)		*Energy investment* — ATP molecules are broken down/used up (to provide energy) OR Phosphorylation/addition of phosphate to glucose/intermediates occur. *Energy pay-off* — ATP molecules are produced	2

Question			Expected answer(s)	Max mark
	(b)		Acetyl (group)/acetyl coenzyme A/ acetyl CoA produced when oxygen is present/in aerobic conditions OR Lactic acid produced when oxygen is absent/insufficient/in anaerobic conditions	2
	(c)		Athlete — Sprinter Reason — creatine (phosphate) releases energy at a fast rate/for a short period of time/runs out quickly.	1
5.	(a)		SS	1
	(b)		50	1
	(c)	(i)	It alters the (DNA) nucleotide sequence OR replaces one nucleotide with another	1
		(ii)	An incorrect amino acid is placed in the protein/polypeptide chain/haemoglobin OR One amino acid is replaced by another in the protein/polypeptide chain/ haemoglobin OR The amino acid sequence is shortened (due to a stop codon)	1
	(d)		Pre-implantation Genetic Diagnosis/ PGD/pre-implantation genetic screening	1
	(e)		This drug could switch on the gene for fetal haemoglobin (in the child so haemoglobin is produced) OR This drug could stop the gene being switched off (in the child)	1
6.	(a)		It can contract/vasoconstrict to reduce blood flow to some areas OR It can relax/vasodilate to increase blood flow to some areas	1
	(b)		1. Endothelium is damaged 2. Clotting factors are released 3. Prothrombin (enzyme) is converted/ activated/changed into thrombin 4. Fibrinogen is converted into fibrin (by thrombin) 5. Fibrin/threads form a meshwork (that seals the wound) 6. The clot/thrombus formed may break loose, forming an embolus 7. A clot/thrombus may lead to a heart attack/stroke	5
7.	(a)	(i)	98 beats/minute	1
		(ii)	Stroke volume increased as oxygen uptake increased, until 2 litres/min, after which it remained constant.	1
		(iii)	150	1
	(b)		18·72	1

Question			Expected answer(s)	Max mark
	(c)	(i)	The first figure is systolic blood pressure/when blood is surging through the arteries/when the artery wall is stretched and the second figure is diastolic blood pressure/when blood is not surging through the arteries/when the artery wall has recoiled	1
		(ii)	High blood pressure forces more fluid out of the capillaries Lymph vessels cannot reabsorb all the excess tissue fluid	2
8.	(a)	(i)	A — Diabetic because blood glucose concentration increases faster/to a higher level/for a longer time OR because blood glucose concentration does not return to normal (after 150 minutes) B — Non-diabetic because blood glucose increases slower/to a lower level/for a shorter time OR because blood glucose concentration returns to normal (after 60 minutes)	1
		(ii)	Blood glucose concentration increases for 60 minutes and then decreases At least one blood glucose concentration given with units eg Start = 4·8 m mol/litre 60 minutes = 11·2 m mol/litre 150 minutes = 7·6 m mol/litre	2
	(b)		Type 1 — Insulin is not produced so blood glucose concentration cannot be controlled Type 2 — Insulin is produced but cells are less sensitive to insulin/have fewer insulin receptors/have developed insulin resistance	2
9.	(a)	(i)	1. As age increases, the frequency/ number of cases of obesity increases 2. The frequency/number of cases is higher in 2012 (compared to 2003)	2
		(ii)	1·536 million/1 536 000	1
	(b)		Reduce their intake of fats/sugars/ carbohydrates OR exercise more/become more active	1
10.	(a)		85·5	1

Question			Expected answer(s)	Max mark
	(b)		1. Each group has a similar gender balance 2. Each group completed the same jigsaw puzzle 3. Each group contained children with similar (physical/mental) abilities 4. The investigation was carried out in the same environmental conditions/same room/same temperature/same time of day/no distractions were present	2
	(c)		As children get older they <u>learn</u> faster (how to complete puzzles)	1
	(d)	(i)	By the fifth attempt the children had learned/memorised where the pieces went (as a result of experience)	1
		(ii)	Some children had become bored with/lost interest in the puzzle (by the fifth attempt/through lack of reinforcement)	1
	(e)		Repeat the investigation in front of an audience/as a competition	1
11.	(a)	(i)	1955 or 1956	1
		(ii)	Decrease in vaccination rate/lack of vaccines available OR mass immigration OR mutation of the whooping cough bacteria OR adverse publicity about the vaccine	1
	(b)		A large percentage of the population have been immunised This means that there is a very low chance that non-immune individuals will come into contact with infected individuals	2
12.	(a)	(i)	Shorter life span/lower survival rate, so no time to develop heart disease	1
		(ii)	Better medical care/more doctors/more hospitals/more drugs OR more use of insecticides/vector control OR clean water/sewage treatment	1
	(b)	(i)	20%	1
		(ii)	300 000	1
13.	(a)		Inhaled air/droplet infection.	1
	(b)	(i)	1986–1991	1
		(ii)	Increased vaccination OR more effective antibiotic treatment	1
		(iii)	Cases of pulmonary TB decreased between 1991 and 2006 <u>while</u> cases of non-pulmonary TB increased between 1991 and 2006	1
		(iv)	11 : 5	1

Question			Expected answer(s)	Max mark
	(c)		HIV attacks <u>lymphocytes</u> reducing the ability of the immune system to respond to the bacterial infection	1
14.	A		1. ANS works automatically/without conscious control 2. Impulses originate in the <u>medulla</u> (region of the brain) 3. It is made up of the sympathetic <u>and</u> parasympathetic systems 4. These two systems are <u>antagonistic</u> in action 5. The sympathetic system prepares the body for fight or flight 6. The parasympathetic system prepares the body for rest and digest 7. Correct description of the effect of the ANS in controlling heart rate 8. Correct description of the effect of the ANS in controlling breathing rate 9. Correct description of the effect of the ANS in controlling peristalsis 10. Correct description of the effect of the ANS in controlling intestinal secretions	7
	B		1. Neurotransmitters relay messages from nerve to nerve/muscle 2. Gap between them is called the <u>synaptic cleft</u> 3. Neurotransmitters are stored in <u>vesicles</u> 4. Arrival of an impulse causes vesicles to fuse with membrane <u>and</u> release neurotransmitter 5. Neurotransmitters <u>diffuse</u> across the cleft 6. Neurotransmitters bind to <u>receptors</u> 7. Receptors determine whether the signal is excitory or inhibitory 8. Neurotransmitters are removed by enzymes/re-uptake 9. Removal prevents continuous stimulation of post-synaptic neurones 10. Summation of weak stimuli can release enough neurotransmitter to fire an impulse	7

HIGHER HUMAN BIOLOGY 2015

Section 1

Question	Answer	Mark
1.	C	1
2.	C	1
3.	B	1
4.	D	1
5.	D	1
6.	D	1
7.	A	1
8.	D	1
9.	B	1
10.	A	1
11.	B	1
12.	D	1
13.	A	1
14.	A	1
15.	B	1
16.	B	1
17.	A	1
18.	C	1
19.	B	1
20.	C	1

Section 2

Question			Expected answer(s)	Max mark
1.	(a)		(Relatively) unspecialised (cells) OR Capable of (repeated) division OR Can differentiate (into specialised cells) OR Are totipotent.	1
	(b)		Embryonic stem cells/inner cell mass cells can form all cells types/are totipotent or pluripotent while tissue/adult stem cells can only form a limited range of cell types/are multipotent.	1
	(C)	(i)	Tissue/stem cells are cultured/grown (in laboratory/outside body). OR Tissue/stem cells are transplanted/placed into the muscle/tissue/damaged area.	1
		(ii)	Stem cells can be used to study diseases/cancer. OR Stem cells can be used for drug/medicine testing/treatment.	1

Question			Expected answer(s)	Max mark
2.	(a)		Anabolic/synthetic/biosynthetic/synthesis	1
	(b)	(i)	It will contain a different nucleotide/base. OR It will contain a different codon/stop codon.	1
		(ii)	The protein/enzyme/glycogen synthase contains a different amino acid(s).	1
	(c)		Glucose is used up in respiration/to provide energy/ATP and they have no glycogen stores to provide more glucose.	1
	(d)		*Recessive* Disease skips generations/does not appear in every generation. OR Two unaffected/heterozygous/carrier parents can have an affected child. *1 mark* *Sex-linked* **More males** will be affected than females. OR **Affected males** do not pass the allele/condition to their sons. OR **Affected males** can only pass the allele/condition to their daughters. OR **Unaffected males** cannot pass the allele/condition to their daughters. OR Only **affected/carrier females** pass the condition to their sons. *1 mark*	2
3.	(a)		• Volume of yeast suspension/solution/cells. • Concentration of yeast suspension/solution/ number of yeast cells/mass of yeast. • Type/age/source of yeast cells. • Area/size/diameter/volume/thickness/type of gel or dish. • Concentration of nutrients in gel/pH of gel. • Strength or intensity of lamp/use same lamp/distance of lamp. • Temperature of incubator/dishes. • Time for yeast to grow/dishes left in incubator. *Any two*	2
	(b)	(i)	Axes have correct scales and labels *1 mark* Points correctly plotted and line drawn (touching each point). *1 mark*	2
		(ii)	Increasing the exposure (to UV radiation) increases the number of yeast cells/colonies that die/are damaged. OR Increasing the exposure (to UV radiation) decreases the number of yeast cells/colonies that survive.	1

Question			Expected answer(s)	Max mark
		(iii)	Repeat the investigation <u>at each exposure</u> (time). **OR** Repeat the investigation and calculate <u>averages</u>.	1
	(c)	(i)	400	1
		(ii)	The number of yeast cells/colonies at SPF 15 is almost as much as with higher SPF values. **OR** There are more yeast cells/colonies using SPF 15 compared to when using no sunscreen.	1
		(iii)	350 minutes/5 hours 50 minutes	1
4.	(a)		ATP is broken down/used up/converted to ADP. **OR** ATP is put into the reaction.	1
	(b)		It changes the shape/form of the active site (to suit the substrate molecule). **OR** It induces a better fit with the substrate. **OR** It <u>lowers</u> the activation energy.	1
	(c)		Enzyme 1 **OR** Enzyme 3 Explanation: The transfer of phosphate/addition of phosphate (from ATP)	1
	(d)		When there has been a build-up/too much/an increased concentration of <u>fructose-6-phosphate</u>.	1
	(e)		This ensures the cell only <u>produces ATP</u> when required **OR** This ensures that <u>glucose is only used</u> when it is required/conserved.	1
5.	(a)	(i)	Releases/supplies <u>energy</u> (rapidly/at a fast rate). *1 mark* <u>Phosphate</u> (released) is used to convert ADP to ATP/to create ATP. *1 mark*	2
		(ii)	The creatine phosphate supply runs out (after 10 seconds).	1
	(b)		Lactic acid/lactate.	1

Question			Expected answer(s)	Max mark
	(c)		Muscle fibre: Slow twitch Sport: any suitable endurance sport Reasons: • they contract (relatively) slowly • can contract over a (relatively) long period • have many mitochondria • have a large blood supply • rely on aerobic respiration (to generate ATP) • have a high concentration of myoglobin • stores/energy source is mainly fat. Muscle fibre: Fast twitch Sport: any sport requiring bursts of energy Reasons: • they contract (relatively) quickly • can contract over a (relatively) short period • have few mitochondria • have a low blood supply • rely on glycolysis (to generate ATP) • have a low concentration of myoglobin • stores/main energy source is glycogen/creatine phosphate. *Any 3 reasons for 3 marks*	3
6.	(a)	(i)	Cross (centre) is placed in the fetal tissue area.	1
		(ii)	Cells are cultured/allowed to divide (to obtain sufficient cells) *1 mark* (Karyotypes then show) the <u>chromosomes</u> (from the cells). *1 mark*	2
		(iii)	CVS can be carried out earlier (in pregnancy than amniocentesis).	1
	(b)		Biochemical	1
7.	(a)		18	1
	(b)		(As children get older) they eat less 'healthy' food/have a higher fat diet/have more sugar in their diet. **OR** (As children get older) they choose/control what they eat. **OR** (As children get older) they exercise less/carry out less physical activities.	1
	(c)		Weight divided by height <u>squared</u>.	1

Question			Expected answer(s)	Max mark
	(d)		Identification — use celebrities/role models/someone they admire to promote a healthy lifestyle/healthy diet/exercise. *1 mark*	2
			Internalisation — Use adverts/media/parents/reasoned arguments/persuasion to get children to adopt a healthy lifestyle/healthy diet/exercise. *1 mark*	
8.	(a)		Between work levels 1 to 5 the stroke volume increased <u>and</u> then it remained constant between levels 5 and 7. *1 mark*	2
			It increased from 88 cm^3 to 140 cm^3 **OR** it remained constant at 140 cm^3. *1 mark*	
	(b)		102	1
	(c)		19 600	1
	(d)	(i)	4·5	1
		(ii)	120 beats/min	1
		(iii)	23·81/23·8/24	1
9.	(a)		<u>Enough/increased/more/a higher concentration</u> of <u>dopamine/neurotransmitter</u> (is released) to trigger an impulse/reach threshold. **OR** <u>Many/a series</u> of <u>weak stimuli</u> trigger an impulse/reach threshold.	1
	(b)		Provides <u>energy/ATP</u> to make/release neurotransmitter or dopamine/form vesicles/allow vesicles to move/allow vesicles to fuse with the membrane/to reuptake neurotransmitter.	1
	(c)		(Cocaine) blocks/inhibits/acts as an antagonist to the re-uptake <u>proteins</u>. **OR** (Cocaine) prevents the <u>proteins</u> from reabsorbing/taking up/removing dopamine. *1 mark*	2
			Dopamine/neurotransmitter remains in the synapse/at the receptors. **OR** Dopamine/neurotransmitter continues to stimulate receptors/fire impulses. *1 mark*	

Question			Expected answer(s)	Max mark
10.	(a)		Organisation	2
			Related information is grouped together. **OR** Information is put into categories/headings. *1 mark*	
			Elaboration	
			Additional information is given (about each term). **OR** Meaningful information is given (about each term). *1 mark*	
	(b)		<u>Short-term</u> memory/STM has a limited capacity/span/only holds around 7 items of information.	1
	(c)		Cerebrum/cortex	1
11.	(a)	(i)	Release histamine *1 mark*	2
			This causes vasodilation/increased <u>capillary</u> permeability *1 mark* **OR** Release cytokines *1 mark*	
			This leads to an accumulation of phagocytes/the delivery of antimicrobial proteins/clotting elements *1 mark*	
		(ii)	NK cells induce/cause the <u>infected/invaded cell</u> to self-destruct/undergo apoptosis/undergo programmed cell death. *1 mark*	1
	(b)		Phagocytes engulf/capture/digest the bacteria/pathogen. *1 mark*	2
			They display the bacteria's/pathogen's <u>antigens</u> on their surface (activating T-lymphocytes). *1 mark*	
12.	(a)		26 million/26 000 000	1
	(b)		0·6 million/600 000	1
	(c)		5·14/5·1/5	1
	(d)		The steep<u>est</u>/steep<u>er</u> part of the (HIV infected) graph/line was between 1993 and 1995 **OR** The graph/line <u>sharply increases</u> between 1993 and 1995 while the rest of the graph/line <u>more steadily</u> increases.	1

Question			Expected answer(s)	Max mark
13.	(a)	(i)	Each group should contain individuals of similar ages/a similar age range. **OR** Each group should contain the same number of males and females/same gender mix. **OR** Individuals in groups should have no recent history of influenza. **OR** Individuals should not be allowed to travel abroad during the study **OR** Individuals should all be in good general health.	1
		(ii)	Did not develop influenza = 453 <u>and</u> Total = 470	1
		(iii)	R	1
	(b)		It enhances/improves the immune response/antibody production. **OR** It improves the <u>effectiveness</u> of the vaccine.	1
	(c)		A pandemic	1
14.	(a)	(i)	1. Pituitary gland secretes/produces FSH/LH. 2. FSH stimulates growth of <u>follicle</u> (in the ovary). 3. Follicle/ovary produces oestrogen. 4. Oestrogen stimulates growth/repair/proliferation/thickening of endometrium/uterus lining. 5. Oestrogen stimulates production of LH. 6. LH (surge) brings about ovulation/release of the egg. 7. <u>Rising/high levels</u> of oestrogen inhibit FSH production. 8. This is negative feedback *Any 6 points for 6 marks*	6

Question			Expected answer(s)	Max mark
		(ii)	a. The follicle develops into the corpus luteum. b. Corpus luteum secretes progesterone (and oestrogen). c. Progesterone maintains/increases/thickens the endometrium/uterus lining. d. Progesterone inhibits <u>FSH/LH</u> production. e. Progesterone/oestrogen levels decrease (towards the end of the cycle). f. This/corpus luteum degeneration triggers menstruation/breakdown of the endometrium. *Any 4 points for 4 marks*	4
	(b)	(i)	1. Pacemaker/SAN contains autorhythmic cells/is where the heart beat originates/is found in the right atrium. 2. <u>Impulse/wave of excitation</u> spreads across the atria/cause the atria to contract/cause atrial systole. 3. (Impulses) reach/stimulate the atrioventricular node/AVN. 4. AVN found at junction of atria and ventricles/at base of atria. 5. Impulses from AVN spread through ventricles. 6. (Cause) contraction of ventricles/ventricular systole. 7. (This is followed by) relaxation/resting/diastolic phase/diastole. *Any 5 points for 5 marks*	5
		(ii)	a. <u>Medulla</u> controls the cardiac cycle/regulates the SAN. b. <u>Autonomic nervous system</u> (carries impulses to heart). c. <u>Sympathetic</u> nerve speeds up the heart rate. d. Sympathetic nerve releases noradrenaline/norepinephrine. e. <u>Parasympathetic</u> nerve slows down the heart rate. f. Parasympathetic nerve releases acetylcholine. g. Sympathetic and parasympathetic systems are <u>antagonistic</u> to each other *Any 5 points for 5 marks*	5

HIGHER HUMAN BIOLOGY 2016

Section 1

Question	Answer	Mark
1.	C	1
2.	D	1
3.	C	1
4.	B	1
5.	D	1
6.	C	1
7.	B	1
8.	A	1
9.	B	1
10.	A	1
11.	C	1
12.	B	1
13.	A	1
14.	D	1
15.	C	1
16.	A	1
17.	B	1
18.	D	1
19.	C	1
20.	B	1

Section 2

Question			Expected answer(s)	Max Mark
1.	(a)	(i)	Hydrogen	1
		(ii)	Phosphate	1
	(b)		<u>Nucleotides</u> can only be added to the 3′/deoxyribose end (of a new strand/ primer). **OR** DNA/strands can only be replicated from 5′ to 3′.	1
	(c)		DNA polymerase adds <u>nucleotides</u> (to the new strand/primer) **(1)** Ligase joins fragments (of DNA/lagging strand) **(1)**	2
2.	(a)	(i)	GUG CUA GCU AUC CUA	1
		(ii)	Exons	1
		(iii)	(RNA/alternative) splicing	1
		(iv)	G C U A	1
	(b)		Ribosome	1
	(c)		Polypeptide chains can be cut/cleaved (and recombined). **OR** Phosphate/carbohydrate groups may be added (to the polypeptide).	1
3.	(a)	(i)	742	1

Question			Expected answer(s)	Max Mark
		(ii)	Between 0 and 7 years after diagnosis the percentage of surviving patients decreased <u>and</u> then it remained constant between 7 and 9 years. **(1)** It decreased from 100% to 62%. **OR** It decreased by 38%. **OR** It remained constant at 62%. **(1)**	2
	(b)		1. (Cancer) cells divide excessively <u>and</u> this leads to a mass of abnormal cells/tumour. 2. These cells don't respond to <u>regulatory signals</u>. 3. <u>Cells</u> fail to attach to each other/ the tumour. **OR** <u>Cells</u> detach from each other/the tumour. 4. They/cells spread to form <u>secondary tumours</u>. **(Any 3)**	3
4.	(a)		1. Volume of solution. 2. Concentration of solution. 3. Initial pH of solution. 4. Diameter of tube/length of tube/ size of tube/position of tube in indicator/size of bottle. 5. Temperature of air/bottle/solution. **(Any 2)**	2
	(b)		40	1
	(c)		They repeated the investigation/ experiment and took an average/ averages. **OR** They repeated the investigation/ experiment at each activity.	1
	(d)		Correct scale on vertical axis and correct labels on both axes. **(1)** Bars are correctly drawn. (33, 25 and 18) **(1)**	2
	(e)		Increasing physical activity increases the <u>respiration rate.</u> **OR** Running produces the highest <u>respiration rate.</u>	1
	(f)		Increased respiration/activity produces more carbon dioxide. **OR** Increased exercise/physical activity uses more ATP/energy. **OR** Running produces the most carbon dioxide/uses the most ATP (or opposite for resting).	1
5.	(a)		Label correctly showing pulmonary artery.	1
	(b)		The blood would contain a lower concentration of oxygen/less oxygen. **(1)** Deoxygenated blood enters the <u>left</u> ventricle/side of the heart. **(1)**	2

Question			Expected answer(s)	Max Mark
	(c)	(i)	Right atrium.	1
		(ii)	Electrocardiogram/ECG.	1
	(d)		They have a higher <u>heart rate</u>.	1
6.	(a)	(i)	28	1
		(ii)	It allows groups <u>of different sizes</u> to be compared. **OR** It allows <u>different populations</u> to be compared.	1
		(iii)	4284	1
		(iv)	3:2	1
	(b)		A blockage/clot/thrombus/embolism in an artery/blood vessel leading to/in the brain.	1
	(c)		Lack of oxygen kills (brain) <u>cells</u>/<u>tissues</u>. Stroke/damage/lack of oxygen occurs in the <u>right side</u>/<u>hemisphere</u> of the brain/motorcortex/cerebrum. Impulses/signals are not transmitted/sent to the muscles (on the left side of the body). **OR** Impulses/signals are not transmitted/sent so preventing movement (on the left side of the body). **(Any 2 from 3)**	2
7.	(a)	(i)	26 or 26·0	1
		(ii)	Their BMI is greater than <u>30</u>.	1
	(b)		Exercise increases energy expenditure/increases respiration rate/uses up (stored) fats.	1
	(c)		They have a (relatively) high muscle mass.	1
8.	(a)		Recommended — it lowers cholesterol levels. **(1)** Not recommended — it takes a long time to work. **(1)**	2
	(b)		A placebo **OR** A capsule containing no statin/no drug.	1
	(c)		**Randomised —** All individuals have an equal chance of being in either group. **OR** Example describing this. **(1)** **Double-blind —** Neither the participants/patients or the researchers/ doctors should know which group participants are placed into/who is getting the drug. **(1)**	2
	(d)		The error bars overlap. **OR** There is no <u>significant</u> difference between the group results.	1
	(e)		Found in cell membranes. **OR** Forms hormones/forms (other) steroids/is a precursor for steroids (being synthesised).	1

Question			Expected answer(s)	Max Mark
9.	(a)		Cerebrum/cerebral hemisphere(s)/cerebral cortex.	1
	(b)		Different/some areas of the brain are used/active during different aspects/parts of the task.	1
	(c)		These areas are receiving signals/impulses from eyes/ears. **OR** These are the hearing/auditory/visual areas. **OR** These are areas where sounds/language/images are processed.	1
	(d)		Sensory area — individual was touching/feeling (the paper). **(1)** Motor area — individual was using (muscles in) hands/fingers (to fold the paper). **(1)**	2
10.	(a)		Axon	1
	(b)		It attaches to a <u>receptor</u>/diffuses into the <u>receptor</u> (on the postsynaptic membrane).	1
	(c)		Fast twitch.	1
	(d)	(i)	Agonists (bind to and) stimulate (neurotransmitter) <u>receptors</u>. **OR** Agonists mimic (the action of) <u>neurotransmitters</u>.	1
		(ii)	Nicotine triggers/causes the (increased) release of/activates <u>dopamine</u>/<u>endorphins</u>. **OR** Nicotine acts as an agonist of/mimics <u>dopamine</u>. **OR** Nicotine stimulates/reinforces the <u>reward</u> pathway/circuit. **OR** Nicotine blocks/prevents/inhibits the reuptake of dopamine.	1
11.	(a)		3·1 million/3 100 000	1
	(b)		30	1
	(c)		70	1
	(d)	(i)	They have a very low chance of coming into contact with/being exposed to someone who has measles/the disease/is carrying the pathogen.	1
		(ii)	Malnutrition/poverty/rejection by some of the population/lack of education/lack of access to medical resources or vaccines/geographical remoteness.	1
	(e)		1242	1
	(f)		If the rate of decrease in the number of measles cases remains the same there will be no cases of measles (by 2020). **OR** Between 2005 and 2010 the number of cases decreased by 300 000. This suggests that measles will be eliminated (by 2015/2020).	1

Question			Expected answer(s)	Max Mark
12.	(a)	(i)	The <u>receptors</u> on the lymphocyte bind to the <u>antigen</u> (on the pathogen). **(1)** This leads to (repeated) division (of the lymphocyte to form a clone). **(1)**	2
		(ii)	Phagocytes capture/engulf pathogens/ bacteria/viruses <u>and display antigens</u>/ become <u>antigen-presenting</u> cells. **(1)** These activate/stimulate <u>T-lymphocytes.</u> **OR** These cause the production of <u>T-lymphocytes.</u> **(1)**	2
	(b)		*Allergy* B lymphocyte. **(1)** Attack/respond to a <u>harmless antigen.</u> **(1)** **OR** *Autoimmune disease* T lymphocyte. **(1)** Attack/respond to <u>self-antigens</u>. **(1)**	2
13.	A		<u>Causes</u> 1. Too much fat/cholesterol in the diet/blood. **(1)** 2. High LDL levels/low HDL levels **or** High LDL:HDL **or** Low HDL:LDL. **(1)** 3. Lack of exercise/inactive lifestyle. **(1)** 4. Genetic condition/familial hypercholesterolaemia/FH. **(1)** 5. Diabetes/high blood glucose levels. **(1)** <u>Development</u> 6. There is an accumulation of fatty/ fibrous material/cholesterol/ calcium. **(1)** 7. The atheroma/plaque forms beneath the <u>endothelium</u> of an artery. **(1)** 8. Artery (wall) thickens/lumen narrows. **(1)** 9. Blood flow is reduced/ restricted/prevented. **(1)** 10. Loss of elasticity in artery (wall)/hardening of the arteries occurs. **(1)** <u>Health Problems</u> 11. Raises blood pressure/causes hypertension. **(1)** 12. Causes CHD/angina/ heart attack/ stroke/PVD. **(any 2)** **(1)** 13. Description of CHD/angina/heart attack/stroke/PVD. **(1)** *In order to score 8 marks candidate must mention at least one point from each of the three areas.*	8

Question		Expected answer(s)	Max Mark
	B	<u>Diagnosis</u> 1. Glucose presence in urine suggests diabetes. **(1)** 2. (Diagnosis made by carrying out a) <u>glucose tolerance test.</u> **(1)** 3. Individual fasts/does not eat prior to the test. **(1)** 4. Individual drinks a glucose solution/ drink. **(1)** 5. Blood glucose concentration that remains high indicates diabetes. **(1)** 6. Type 1 diabetes tends to be diagnosed in children <u>while</u> type 2 diabetes tends to be diagnosed in adults/later in life. **(1)** <u>Treatment</u> 7. Type 1 diabetes is treated with regular doses/injections of insulin. **(1)** 8. Type 2 diabetes is treated/ controlled by lifestyle changes/ weight loss/exercise/dietary changes. **(1)** <u>Role of Insulin</u> 9. Insulin is produced in the pancreas. **(1)** 10. Type 1 diabetics are unable to produce insulin. **(1)** 11. Insulin converts glucose into glycogen. **(1)** 12. Type 2 diabetics can produce insulin but cells are less sensitive/resistant to it. **(1)** 13. In type 2 diabetics there are fewer insulin receptors on cells. **(1)** *In order to score 8 marks candidate must mention at least one point from each of the three areas.*	8